Ecology, Cognition and Landscape

Landscape Series

Volume 11

Series Editors:

Henri Décamps
Centre National de la Recherche Scientifique
Toulouse, France
Bärbel Tress
TRESS & TRESS GbR
Munich, Germany
Gunther Tress
TRESS & TRESS GbR
Munich, Germany

Aims and Scope

Springer's innovative Landscape Series is committed to publishing high-quality manuscripts that approach the concept of landscape from a broad range of perspectives. Encouraging contributions on theory development, as well as more applied studies, the series attracts outstanding research from the natural and social sciences, and from the humanities and the arts. It also provides a leading forum for publications from interdisciplinary and transdisciplinary teams.

Drawing on, and synthesising, this integrative approach the Springer Landscape Series aims to add new and innovative insights into the multidimensional nature of landscapes. Landscapes provide homes and livelihoods to diverse peoples; they house historic – and prehistoric – artefacts; and they comprise complex physical, chemical and biological systems. They are also shaped and governed by human societies who base their existence on the use of the natural resources; people enjoy the aesthetic qualities and recreational facilities of landscapes, and people design new landscapes.

As interested in identifying best practice as it is in progressing landscape theory, the Landscape Series particularly welcomes problem-solving approaches and contributions to landscape management and planning. The ultimate goal is to facilitate both the application of landscape research to practice, and the feedback from practice into research.

For other titles published in this series, go to
www.springer.com/series/6211

Almo Farina

Ecology, Cognition and Landscape

Linking Natural and Social Systems

 Springer

Prof. Almo Farina
Urbino University
Department of Mathematics, Physics
and Informatics
Campus Scientifico Sogesta
61029 Urbino, Italy
almo.farina@uniurb.it

ISBN 978-94-007-3081-6   e-ISBN 978-90-481-3138-9
DOI 10.1007/978-90-481-3138-9
Springer Dordrecht Heidelberg London New York

Printed on acid-free paper

Springer is part of Springer Science+Business Media (www.springer.com)

Contents

Introduction

It is more and more evident that our living system is completely disturbed by human intrusion. Such intrusion affects the functioning of entire systems in ways we do not yet fully understand. We use paradigms such as the disturbance to cover large and deep gaps in our scientific knowledge.

Human ecology is an uncertain terrain for anthropologists, geographers, and ecologists and rarely is expanded to include the social and economic realms. The integration of different disciplines and the application of their many paradigms to problems of environmental complexity remains a distant goal despite the many efforts that have been made to achieve it. Philosophical and semantic barriers are erected when such integration is pursued by pioneering scientists.

Recently, evolutionary ecology has shown great interest in the spatial processes well described by the emerging discipline of landscape ecology. But this interest takes the form of pure curiosity or at worst, of skepticism toward the real capacity of landscape ecology to contribute to the advancement of ecological science.

The past two centuries have been characterized by huge changes occurring in the entire ecosphere. Global changes are the effects of human intervention at a planetary scale, with consequent degradation of the environment creating an ecological debt for future generations. On the other side of the issue, new technologies have improved the welfare of billions of people and have given hope to many other billions that they may also see such improvement in the near future.

New economics, faster movement of populations, resource deterioration, and global nets of rapidly moving information are some of the relevant effects of this time.

Recently, J. Lawton invoked the birth of a new science known as Earth System Science. I agree with this idea, although not on the title, to develop a more integrated science able to link together different processes including those from economics and politics.

Human societies are facing structural and organizational changes at an increased rate and demand new visions of these changing realities. We must become able to anticipate future scenarios, and to arrange new and more efficient tools to reduce, compensate, and remediate the environmental deterioration.

The present time is characterized by societies more and more implicated in the forceful management of natural resources and by the disappearance of natural as well as cultural values.

For this reason it is urgent to find new tools able to detect the everyday modifications of our living support system. Global changes modify patterns, processes, and species diversity. These changes occur along a broad range of spatial and temporal scales and across many relevant processes. It is urgent to determine where in the real world such changes occur. Such places in the real world become hot spots, or areas where genetic variability is high and where organisms perceive their biological limits.

For this and many other reasons I introduce, adopting different perspectives, landscape ecology as a new framework from which to investigate and speculate on environmental complexity and on the constraints regulating natural dynamics. The goal is to create a bridge from ecosystem ecology to landscape ecology and the complexity of the real world dominated by humans.

For many years the landscape has been considered both a reference concept and a place to develop metrics useful for the management of natural resources and, more generally, to understand the human interaction with the environment. The landscape can be considered as an entity emerging from the interaction of different species and their integrated biophysical processes.

Many contributions have discussed whether the landscape is a subject to be addressed in a multidisciplinary or transdisciplinary way. I strongly believe that to study the landscape we need a new science. This is the main aim of this book and my final goal is to demonstrate that a robust theoretical basis is required to make significant progress in this field.

It is difficult to abandon well recognized and popular paradigms and to venture in unknown directions at the frontiers of a discipline. It is with this feeling that I embark upon this new book.

In this short book I'll try to approach the landscape quite differently, and will propose new ideas, concepts, and possible applications in order to move out of the stagnation into which landscape ecology has fallen.

This book's main priority is to develop a robust theory on the landscape, proposing new ideas and revising the influential literature produced since Carl Troll introduced the first foundations of landscape ecology in the first half of the 20th century.

In particular two theories are discussed: the general theory of resources and the mosaic theory. The first considers the landscape a semiotic interface individual specific, the second recognized the role of the (land) mosaic as a major driver of ecosystemic processes. The two visions are not in competition but represent two ways of approaching the complexity of the real world and to clearly define the spatial dimension.

Both theories require several paradigms and distinct tools and may receive different levels of attention according to the cultural background of the scholars.

Adopting a simple metaphor we could say that the general resources theory describes the organism's perception and cognition of the world with the feet poised firmly on the land. The mosaic theory describes a world observed from an orthogonal distance from the soil. The representation of the reality is based on an aerial perspective and shape and size (patterns) of spatial objects act as major drivers of the ecological processes.

Urbino, Italy   Almo Farina

Chapter 1
The State of Art of Landscape Ecology:
20 Years of Paradigms and Methods

Introduction

In complete agreement with the thoughts of Kuhn (1962), scientific research similarly to other human activities provides a service that is intimately associated with
time, place, and culture. In other words, scientific research intercepts the needs of
human society and tries to describe the related processes. In particular over the last
few years the landscape paradigm has become more and more popular. But it seems
to the author an impossible mission to describe and monitor all the changes that
have occurred along the geographical land zones around the world. As elucidated
by Robert MacArthur (1972), every place on our Planet has a specific ecology and
if we exclude the organismic teleonomy of the biological components (virus, bacteria, plants, and animals) that react to general metabolic allometric rules (Brown
et al. 2004), ecosystems and landscapes appear to be systems functioning under
local constraints and surrounding stochasticity.

In the time of MacArthur (1960s and 1970s) the landscape was not recognized
as an ecological entity but only as a geographical entity.

Today it is recognized that a landscape is the result of meta-ecosystemic processes coupled with cognitive ones, where energy, information, and cybernetic
mechanisms are interacting and integrating to produce emergent patterns (mosaics)
and processes (resource-oriented suitability).

On several occasions ecologists have categorized problems, needs, and environmental priorities associated with indicators, actions, and recommendations (see
Lubchenco et al. 1991). Commendable has been the effort made by Rapport
(Rapport et al. 1998) to consider the ecosystem as a unitarian entity introducing
the concept of "ecosystem health."

But despite the huge amount of articles and special issues that have appeared in
important magazines like Nature and Science the ecology of landscapes in a global
scenario has had a negligible impact (see Myers et al. 2000).

Today financial and economic mechanisms seem to be the major actors able to
modify the functions and speed of the Earth's gears. The recent worldwide financial crisis has cascade effects on most countries in the world and is perceived by
the population as more dramatic than climatic changes. Probably this is related to

A. Farina, *Ecology, Cognition and Landscape*, Landscape Series 11,
DOI 10.1007/978-90-481-3138-9_1, © Springer Science+Business Media B.V. 2010

the contemporary of the crisis when compared to global changes that have local diachronic effects.

Unlike economy, ecology is a science that studies natural machines and does not have operational tools able to modify such a project but only educational tools such as "The Millennium Ecosystem Assessment" to inform about the consequences of ecosystem change for human well-being, alerting people to present and future damage observed in natural modified ecosystems.

From the Ecosystem Concept to the Landscape: Historical and Scientific Motives

Since the English ecologist Sir Arthur Tansley introduced the paradigm of the ecosystem at the beginning of the past century, ecology has been projected into the human realm.

The evolution of ecological thinking has not always been easy and is presently quite incomplete as well described by Frank Golley in "A History of the Ecosystem Concept in Ecology" (Golley 1993).

A significant role in the development of the ecosystem concept was played by Raymond Laurel Lindeman with his innovative and inspirational work in the PhD thesis titled "Cedar Bog Lake: The Ecosystem or the Trophic Dynamic-Viewpoint in Ecology."

For more than 50 years the ecosystem paradigm has been at the center of ecological thinking connecting the nonbiotic world (matter, energy, and statistical and physical information) with the biological world (living processes) as well emphasized for instance by Oward Odum in "System Ecology" (Odum 1983) and Ramon Margalef in "Our Biosphere" (Margalef 1997).

In the past 50 years technology has provided humanity with new tools able to manipulate energy and grant the ability to access unexploited resources, but this fact has reduced the scalar interaction between humanity and nature. For instance, the significant use of fossil and nuclear energy is producing climatic changes on a global scale and economic crisis.

Following this critical condition social discomfort is felt either by the population of the richest countries due to a life style based on profit or by the poor countries contaminated by un-sustainable societal and economic processes. Natural disasters and economic crises seem to result from the same process. In vain ecology has tried to inform and alert decision makers and central governments of the risk of dramatic instability in ecosystems and human societies embedded in a self-organizing complexity.

Since Ludwig von Bertalanffy published the controversial book "General System Theory" in 1969 (von Bertalanffy 1969) in which new ideas were proposed about the functioning of systems, the paradigm of complexity appears as a bulwark for a modern ecology not fully aware of the necessity to reduce entropy in human development (Levin 1999).

Two distinct trends can be observed in ecology today. The first is represented by research working in the direction of behavior of molecular entities. The second considers macro processes and their emergent properties (Morowitz 2002), embedded in a foggy context often branded as vitalism in which emerges the chaotic nature of several phenomena (Kauffman 1993, Prigogine 1993, Cushing et al. 2003).

After the emergence of the paradigm of complexity and the possibilities of its application to the biological realm ecology entered a new phase. Contemporarily this idea is accompanied by an emergent paradigm of spatial ecology, called landscape ecology, which creates an explicit context in which ecosystems and their associated complexity interact (Turner 2005).

In a few decennia landscape ecology has grown in importance and scientific reputation (Naveh 2000, Farina 2006) thanks to the contribution of ecologists, geographers, architects, agronomists, environmental psychologists, and anthropologists pooled by the common necessity to create a context for natural and human processes (Nassauer 1997, Ingold 2000, Wu and Hobbs 2007, Forman 2008).

Brief History of Landscape Ecology

Landscape ecology is a young but well-recognized ecological discipline dealing with the spatial distribution of organisms, patterns, and processes. This discipline developed after the Second World War in central and eastern Europe as an applied science used to manage the countryside. It became popular as a basic science, especially in the US, only during the last two decades.

The theoretical and empirical bodies of this discipline are growing fast but not in a unified fashion (Wu and Hobbs 2002, Metzger 2008). A long and intense debate between ecologists has accompanied its development, but without yielding a concrete agreement about the disciplinary range, the basic paradigms, and its relationship with the dated ecosystems approach. Despite this uncertainty, landscape ecology has attracted several students from many different disciplines including geography, biology, engineering, planning and land management, countryside conservation and, more recently, economics. Ecologists and spatial modelers entered into this "cultural arena" only recently (INTECOL Congress of Syracuse, NY, 1986).

Three different themes can be recognized in landscape ecology during its evolution (Farina 1993) (Fig. 1.1). The first theme was inspired by the patterned complexity or the simplification introduced into the environment by human use. This approach considers the landscape as a mosaic of patches of forested, cultivated, and urbanized areas. According to this vision, humans are responsible for most of the land modifications. The active role of humans as principal modifiers is a central part of the research.

The second theme emerged in the US and deals with the ecology of large areas (landscapes). Such an approach seems extremely important for managing the remaining areas in which human disturbance has not been great or in which natural

Fig. 1.1 Three different
visions of landscape ecology
and major regions in which
the different approaches have
been developed (Farina 1993)

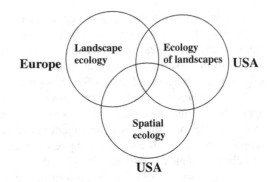

processes are persisting in spite of continuous human development. Nature conservation in natural parks seems to be one of the major areas in which landscape
ecology could be an effective tool for forecasting the changes both inside and outside
such areas.

Broad-scale processes such as erosion or fire can be studied using a plethora of
tools that span the application of remote sensing techniques, GIS, spatial metrics,
and spatial statistics.

The third theme takes into consideration the processes that are dominated by a
spatial context, particularly, the spatial arrangement of organisms in a matrix. Such
an approach is very promising and attractive from many points of view. The spatial
arrangement of organisms indicates the distribution of the resources, and secondly
describes the relationships between and among populations and species. Although
not yet well integrated into ecological disciplines, spatial ecology has favored the
development of many new ideas and paradigms and has allowed conjunction under
the roof of a robust framework of sparse but not for this reason less relevant theories
(e.g. Island Biogeography, MacArthur and Wilson 1967; Meta-population theory,
Hanski and Gilpin 1997), paradigms (e.g. Ecological complexity, Levin 1999) and
models (e.g. source-sink model, Pulliam 1988).

For each vision it is necessary to adopt specific paradigms. The European landscape ecology approach requires implementation of social and economical human
models into the landscape domain. The ecology of the landscapes investigates the
functioning of the mosaics and requires a robust theory of mosaics (see Chapter 2).
Finally, spatial ecology considers the effects of shape and spatial arrangement of
mosaic components on the behavior and on the physiology of plants and animals.

Landscape ecology for many years was considered a discipline quite remote
from the ecological realm, but today the reduced hostility of traditional ecologists, has allowed the incorporation of contributions from landscape ecology into
the canonical sessions of the most important congresses of ecology.

Unfortunately most ecology textbooks don't reference landscape arguments,
although recently some examples have begun to appear (Dodson et al. 1998). The
reason for such diffidence is based on at least two different problems. The first is
the development of a theoretical basis for landscape without taking into account the

paradigms that form the basis of traditional ecology. It seems that landscape ecologists have elaborated new ideas and paradigms starting from a different perspective than that in which ecology is rooted. I suspect that human processes have been so far outside ecological arguments for so long that the sudden recognition of such a plethora of topics bypasses the "cultural niche" and "cage" in which ecology has been developed.

A second problem is due mainly to misuse of landscape ecology's paradigmatic and theoretical approach in technical procedures for land evaluation and planning. Often environmental practitioners and landscape architects have extensively used landscape concepts, procedures, and metrics without the necessary experimental or empirical verification, assuming the exactness of tools not fully validated. This is particularly evident in the quantitative approach to the study of landscapes. Although many metrics are now available to measure the regularities/irregularities of landscape shape, their behavior is not fully explained by the processes that we intend to evaluate (e.g. Bogaert 2000, Bogaert et al. 2002).

I argue often that in the "real world," and outside the academic microcosm, efficient approaches to manage and solve problems are urgently required. This has been especially true during the past few decades in which the technological revolution has extended from agriculture to industry, and to all functioning of society. It has strongly impacted the environment, displacing processes and creating pathologies in many functional segments at a broad range of spatial and temporal scales.

One criticism cited against landscape ecology is based on the "superficiality" by which we give credit to spatial patterns. For instance, the concept of the corridor is abused for many purposes when not connected with a specific process or organismic life trait, and over-evaluating the role of such structures often confuses patterns with processes.

Other examples are from the common practice to use thematic maps to organize protected areas or other focal management areas. The world that we represent on maps is perceived differently by different organisms, and this creates a discrepancy between the objects observed by using human perception and their role in the environmental realm.

Finally, we are assuming a much too simplified vision of environmental complexity. It seems landscapes approach an intermediate phase from a functional perception of the environment (ecosystemic ecology) to the patterned "geographic" ecology, although fractal geometry (Mandelbrot 1977) and macroecology (Brown and Maurer 1989, Brown 1995) have opened new perspectives.

In my opinion, and this is the reason for this book, there should exist other levels in which the environment organizes function and patterns. To discover such levels it is necessary to create and use new paradigms and to formulate new theories. Another important task consists of fighting against the conservative reactions of "normal science" practitioners who move between auto-ecology and ecosystem science and landscapes without conceptual bridging and integration.

The emergent discipline of landscape ecology in which theoretical and applied fields are recognized (Naveh and Lieberman 1984, Forman and Godron 1986, Forman 1995, Zonneveld 1995), contributes to an integration of ecosystem paradigms with spatial processes across a broad range of spatial and temporal scales (Delcourt and Delcourt 1988), as documented by a rich scientific literature arising around this new theme (Farina 1998, Farina 2006, Dodson et al. 1998) and the numerous debates (Naveh 2007, Antrop 2007).

The convergence toward the landscape includes the perspectives of geographers, biologists, ethologists, ecologists, wildlifers, and landscape architects. The integration of these perspectives has created a productive forum for matching new ideas and approaches and, at the same time, for developing a common framework in which to realize new conceptual syntheses.

Landscape ecology has played a central role in attempting to move a consistent portion of ecology from the past stagnant condition to which this science was relegated, at the end of the 1970s, to accept the challenge to conserve natural resources through more ethical land use and management.

In particular the landscape ecology approach to nature conservation and land management has produced a new impetus in applied sciences, including new conceptualizations of biodiversity (Richtie and Olff 1999, Whittaker 1999) and environmental health (Rapport et al. 1998).

Landscape ecology focuses mainly on patterns and processes scaled according to human perception of the landscape and considered at a spatial level between the ecosystem and the biome (Odum 1989). The study of human interference with natural systems and the possibilities of managing natural and man-made resources in a durable and nondestructive (sustainable) fashion are of special interest.

The ecology of the landscape considers the complexity of ecological systems at a large scale that supersedes the functional scale often associated with the composition of large-scale ecosystems. At this larger scale, many processes across different ecosystem boundaries encounter the heterogeneity of the living substrate, and thus the relationship between the component parts (patches, tesserae, ecotopes), becomes the main goal of the research (Turner 1989, 2005). Such an approach is consistent with the human-perceived landscape and allows the tracking of processes and organisms across regions, watersheds, and territories.

Finally, spatial ecology focuses on a plethora of processes and patterns that are associated with space. Special emphasis is given to the study of animal behavior and to the relationship between habitat structure and organism function. An experimental approach using microcosms is of growing importance in the validation of spatially explicit models that apply statistical tools and fractal mathematics to plants and animals (e.g. Barrett and Peles 1999).

A New Model Links Landscape Ecology to Ecosystem Ecology

Recently, in order to reduce the uncertainties created by the convergence of distant disciplines focusing on environmental complexity, I distinguished two main ontological views: the "process" perspective and the "organismic" perspective (Farina 1998). Such a distinction is not a novelty in ecology, or in the other biological sciences, but it nevertheless provides a useful approach to explore the "black box" of environmental complexity.

The "process" perspective views erosion, fragmentation, movement of organisms or other (disturbance) events as entities that behave according to the different environmental pressures they encounter contributing to landscape heterogeneity. For instance, topography, wind direction and strength, vegetation types, and land use have a great influence on severity and behavior of wild-fire processes.

The second perspective takes into account the "organismic" perception of the environment from a species-specific point of view. For instance, the use of habitat patches for such vital functions as feeding, roosting, breeding, and mating is considered according to species-specific patch suitability and the history of the land mosaic (Turner et al. 1997).

But despite this vision, the uncertain border between the ecosystem, in which the "topological" functions prevail over the "chorological" ones, and the spatial dimension of landscape still persists.

Emerging Properties of the Landscape

Until the present, environmental complexity has not been described by general rules, but by applying a plethora of small, separate rules valid only within a narrow component of such complexity (Allen and Hoekstra 1992). A landscape can be considered as any piece of the "real world" ranging from the few millimeters of a microcosm of pico-plankton, to the several kilometers-wide geographical range occupied by a wolf pack. Objects within a landscape are subject to the functional constraints of emergent properties including heterogeneity and influence from other co-specific, or heterospecific, objects. Density-dependent functions, connectivity, and matrix hostility are some of the emergent properties of a landscape from either a process-based or an organismic-based view.

The landscape paradigm is useful when applied to the habitat concept. A habitat is often not coincident with an ecosystem, but it may be a part of an ecosystem or conversely may be composed of more than one ecosystem. For instance, the habitat of a dragonfly varies greatly from the aquatic larva to the adult that lives in terrestrial habitats. European thrushes overwinter in open agro-ecosystems around the Mediterranean basin, but breed in the dense woodlands of central and northern Europe. In this way, if we consider the functions of a species, we can imagine a habitat that is composed of parts of different ecosystems.

The application of new, more integrated paradigms, such as the eco-field (Laszlo 1996, Farina 2000, Farina and Belgrano 2004, 2006, see also Chapter 8), provides landscape models with new information on the interactions between an organism and its life-support systems. The eco-field can be considered as the functional and structural space in which an organism lives, and it represents the incorporation of the chorological components of the habitat into the niche concept. We predict that an organism crosses a different landscape according to a specific function, integrating the scaling properties of environmental complexity.

Ecological Theories and Models Incorporated into the Landscape Ecology Paradigm

Hierarchy theory (Allen and Starr 1982, O'Neill et al. 1986) is relevant for exploration of the landscape processes. The hierarchy concept is consistent with the structural (patches) and functional (ecotopes) components of a landscape (Fig. 1.2).

Heterogeneity, which is an intrinsic feature of every landscape, can be modeled with the percolation theory of fluids. Such a theory appears useful to evaluate the degree of connectivity of the landscape mosaic as perceived by a species (Gardner et al. 1987). Theoretical landscape ecologists have been interested in this approach for at least the last two decades (e.g. Pearson and Gardner 1997). The combination of neutral models with fractal geometry and Geographic Information Systems has started to clarify the complex responses of organisms and systems to landscape heterogeneity (Milne 1997).

Meta-population theory (Levins 1970) and source-sink models (Pulliam 1988, 1996) seem to be promising frameworks from which to explore, respectively, the effects of habitat fragmentation and differentiated availability of resources. These two paradigms have a common basis in the irregular distribution of habitats and resources. In particular, meta-population theory considers the distribution of populations across fragmented and isolated habitats. The colonization/extinction rate is considered a central mechanism to maintain meta-population health. Gene flow between connected subpopulations reduces the risk of inbreeding and genetic erosion (Hanski and Gilpin 1997, Hanski 1999). Empirical and experimental evidence has quickly consolidated this theory that finds a patterned landscape as the context in which it can be adequately developed (Wiens 1997).

Source-sink models analyze population survival in terms of the balance between reproduction and death within a seasonal cycle. Such models are strictly related to habitat quality and, consequently, to the patterned landscape (Pulliam 1988,

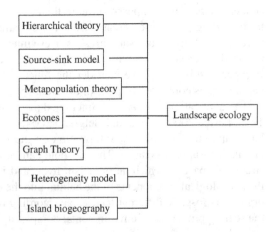

Fig. 1.2 Some relevant theories and models incorporated in the disciplinary body of landscape ecology

1996). Recently, such models have been extended conceptually to estimate the quality of the environment and nonreproductive traits of species, like migration and over-wintering (Farina, unpublished).

To link ecosystems and habitats, the ecotone paradigm seems a very promising approach. Ecotones have received a lot of attention in recent decades, and their importance for energy distribution and the movement/distribution of organisms across a mosaic has been emphasized (Farina 1995). Ecotones are emergent characteristics of mosaic heterogeneity, and exist across spatial and temporal scales. They are species-specific and represent the position in the space-time context in which a particular organism perceives changes in environmental properties.

The ecotone paradigm can describe processes as well as environmental constraints on organisms. The capacity to extend such a paradigm throughout landscape ecology emphasizes the importance of linking ecosystems to functional landscapes. Ecotones represent the border between suitable and unsuitable habitat, delimiting in addition the safe area from the unsafe area. This paradigm is useful for describing pathways between different systems, as well as the degree of hostility of a patch in a habitat matrix (Wiens et al. 1985). Ecotone principles can explain part of the environmental complexity that is realized when different ecosystems are connected, and, consequently, it seems right to measure the ascendency properties of these systems (Ulanowicz 1997). At the same time, the ecotone paradigm can be used to evaluate the complexity of animal movements and their reluctance in crossing different environmental mosaics. Behavioral ecology (Wiens 1999) and cognitive ecology (Real 1993) could contribute to this paradigm.

Concluding Remarks

The landscape is an emergent level of functional complexity. Such a level appears when space and time are extrapolated from an un-scaled functional or structural level. Landscape ecology can describe a part of an ecosystem or a mosaic of inter-related ecosystems. These two visions are not interdependent but they suggest that in nature processes and patterns often appear in a self-similar fashion. Emerging properties at a specific geographical scale and the effect of the composite mosaic are two faces of the same coin. Such emerging characteristics can be interpreted by organisms according to their different life traits. For instance, soil heterogeneity influences vegetation and soil organisms that in turn affect meso and large animals. A common feature of any landscape is its heterogeneity. This heterogeneity may be scaled on structure, water, and nutrients for plants, or on suitable habitat-patches for animals. A matrix in landscape ecology is comprised of the dominant cover of a land mosaic (Forman and Godron 1986) and the physical context of separated patches. Organisms interact with a matrix that is heterogeneous in different ways and at different spatial and temporal resolutions and change the scale to intercept resources (food, water, light, social interactions, etc.).

Every function must find the correct scaled dimension at which to connect the organism with resources. Energy is allocated, removed, or transformed to fulfill such function and feedback mechanisms, which are formally described by ecosystem ecology (Odum 1983, Whittaker 1975). Landscape ecology represents an integrating stage between the consolidated ecosystem paradigm and the "real world" of environmental complexity in space and time.

In conclusion, we could define the landscape as the patterned aspect of complexity. A better description is the environmental context in which biological functions become spatially explicit. This role is fundamental for vegetation ecology as well as animal ecology, and especially in human ecology, but it will require time to be unanimously accepted by researchers and practitioners.

In the following chapters we will develop new ideas to achieve a more precise knowledge of mechanisms that operate in this complex world. Many concepts will be reformulated such as the "domain" paradigm, and others, including the eco-field hypothesis and the mosaic theory, will be revisited and discussed in detail in light of new syntheses.

Landscape ecology has opened a new way to address complexity reducing the gap between the biological and cognitive sciences and between basic research and its application. This effort has opened the road to the formulation of new paradigms and, more generally, to the founding of a new science focused on complexity in which biological processes can be investigated from different perspectives.

Suggested Reading

Allen, T.F.H. and Hoekstra, T.W. 1992. Toward a unified ecology. Columbia University Press, New York.

Barrett, G.W. and Peles, J.D. (eds.) 1999. Landscape ecology of small mammals. Springer-Verlag, New York.

Bastian, O. and Steinhardt, U. (eds.) 2002. Development and perspectives of landscape ecology. Kluwer Academic Publishers, Dordrecht.

Bissonette, J.A. and Storch, I. (eds.) 2003. Landscape ecology and resource management. Linking theory with practice. Island Press, Washington, DC.

Burel, F. and Baudry, J. 2003. Landscape ecology. Concepts, methods and applications. Science Publishers, Enfield, NH.

Farina, A. 2006. Principles and methods in landscape ecology. Springer, Dordrecht.

Forman, R.T.T. and Godron, M. 1986. Landscape ecology. Wiley & Sons, New York.

Forman, R.T.T. 1995. Land mosaics. The ecology of landscapes and regions. Cambridge Academic Press, Cambridge, UK.

Forman, R.T. 2008. Urban regions. Cambridge University Press, Cambridge.

Gergel, S.E. and Turner, M. 2002. Learning landscape ecology. Springer, New York.

Green, D.G., Klomp, N., Rimmington, G., Sadedin, S. (eds.) 2006. Complexity in landscape ecology. Springer, Dordrecht.

Nassauer, J.I. (ed.) 1997. Placing nature. Culture and landscape ecology. Island Press, Washington, DC.

Naveh, Z. and Lieberman, A. 1994. Landscape ecology. Springer-Verlag, New York.

Sanderson, J. and Harris, L.D. 2000. Landscape ecology. A top-down approach. Lewis Publishers, Boca Raton, FL.

Turner, M.G., Gardner, R.H., and O'Neill, R.V. 2001. Landscape ecology in theory and practice. Pattern and process. Springer, New York.

Wu, J. and Hobbs, R.J. (eds.) 2007. Key topics in landscape ecology. Cambridge University Press, Cambridge.

Zonneveld, I.S. and Forman, R.T.T. (eds.) 1990. Changing landscapes: An ecological perspective. Springer-Verlag, New York.

References

Allen, T.F.H. and Starr, T.B. 1982. Hierarchy, perspectives for ecological complexity. The University of Chicago Press, Chicago.

Allen, T.F.H. and Hoekstra, T.W. 1992. Toward a unified ecology. Columbia University Press, New York.

Antrop, M. 2007. Reflecting upon 25 years of landscape ecology. Landscape Ecology 22: 1441–1443.

Barrett, G.W. and Peles, J.D. (eds.) 1999. Landscape ecology of small mammals. Springer-Verlag, New York.

Bogaert, J. 2000. Quantifying habitat fragmentation as a spatial process in a patch-corridor-matrix landscape model. PhD Dissertation, Faculteit Wetenschappen, Departement Biologie, University of Antwerpen.

Bogaert, J., Myneni, R.B., and Knyazikhin, Y. 2002. A mathematical comment on the formulae for the aggregation index and the shape index. Landscape Ecology 17: 87–90.

Brown, J.H. 1995. Macroecology. The University of Chicago Press. Chicago, IL.

Brown, J.H., Gillooly, F., Allen, A.P., Savage, M., and van, West, G.B. 2004. Toward a metabolic theory of ecology. Ecology 85(7):1771–1789.

Brown, J.H. and Maurer, B.A. 1989. Macroecology: The division of food and space among species on continents. Science 243: 1145–1150.

Cushing, J.M., Costantino, R.E., Dennis, B., Desharnais, R.A., and Henson, S.M. 2003. Chaos in ecology. Experimental nonlinear dynamics. Academic Press, San Diego, CA.

Delcourt, H.R. and Delcourt, P.A. 1988. Quaternary landscape ecology: Relevant scales in space and time. Landscape Ecology 2: 23–44.

Dodson, S.I., Allen, T.F.H., Carpenter, S.R., Ives, A.R., Jeanne, R.L., Kitchell, J.F., Langston, N.E., and Turner, M.G. 1998. Ecology. Oxford University Press, New York.

Farina, A. 1993. From global to regional landscape ecology. Landscape Ecology 8(3): 153.

Farina, A. 1995. Ecotoni. Patterns e processi ai margini. CLEUP, Padova.

Farina, A. 1998. Principles and methods in landscape ecology. Chapman & Hall, London.

Farina, A. 2000. Landscape ecology in action. Kluwer Academic Publishers, the Netherlands.

Farina, A. 2006. Principles and methods in landscape ecology. Towards a science of landscape. Springer, Dordrecht.

Farina, A. and Belgrano, A. 2004. The eco-field: A new paradigm for landscape ecology. Ecological Research 19: 107–110.

Farina, A. and Belgrano, A. 2006. The eco-field hypothesis: Toward a cognitive landscape. Landscape Ecology 21: 5–17.

Forman, R.T.T. 1995. Land mosaics. The ecology of landscapes and regions. Cambridge Academic Press, Cambridge, UK.

Forman, R.T.T. and Godron, M. 1986. Landscape ecology. Wiley & Sons, New York.

Gardner, R.H., Milne, B.T., Turner, M.G., and O'Neill, R.V. 1987. Neutral models for the analysis of broad-scale landscape pattern. Landscape Ecology 1: 19–28.

Golley, F.B. 1993. A history of the ecosystem concept in ecology. Yale University Press, New Haven.

Hanski, I. 1999. Metapopulation ecology. Oxford University Press, Oxford, UK.

Hanski, I.A. and Gilpin, M.E. 1997. Metapopulation biology. Academic Press, San Diego, CA.

Ingold, T. 2000. The perception of the environment. Routledge, London.

Kauffman, S. 1993. The origin of order. Oxoford University Press, New York.

Kuhn, T.S. 1962. The structure of scientific revolutions. University of Chicago Press, Chicago, IL.

Laszlo, E. 1996. The whispering pond. Element Books, Inc., Rockport, MA.

Levin, S.A. 1999. Fragile dominion. Complexity and the commons. Helix Books, Perseus Books, Reading, MA.

Levins, R. 1970. Extinction. In: Gertenshaubert, M. (ed.), Some mathematical question in biology. Lectures in mathematics in the life sciences. American Mathematical Society, Providence, Rhode Island, pp. 77–107.

Lubchenco, J., Olson, A.M., Brubaker, L.B., Carpenter, S.R., Holland, M.M., Hubbell, S.P., Levin, S.A., MacMahon, J.A., Matson, P.A., Mellillo, J.M., Mooney, R.A., Peterson, C.H., Pulliam, H.R., Real, L.A., Regal, P.J., Risser, P.G. 1991. The sustainable biosphere initiative: an ecological research agenda. Ecology 72(2):371–412.

MacArthur, R. 1972. Geographical ecology: Patterns in the distribution of species. Harper & Rowe, New York.

MacArthur, R.H. and Wilson, E.O.1967. The theory of island biogeography. Princeton University Press, Princeton, NJ.

Mandelbrot, B.B. 1977. The fractal geometry of nature. Freeman, New York.

Margalef, R. 1997. Our Biosphere. Ecology Institute, Oldendorf/Luhe, Germany.

Metzger, J.P. 2008. Landscape ecology: Perspectives based on the 2007 IALE world congress. Landscape Ecology 23: 501–504.

Milne, B.T. 1997. Application of fractal geometry in wildlife biology. In: Bissonnette, J.A. (ed.), Wildlife and landscape ecology. Effects of pattern and scale. Springer-Verlag, New York, pp. 32–69.

Morowitz, H.J. 2002. The emergence of everything: How the world became complex. Oxford University Press, New York.

Myers, N., Mittermeier, R.A., Mittermeier, C.G., da Fonseca, G.A.B., Kent, J. 2000. Biodiversity hotspots for conservation priorities. Nature 403:853–858.

Nassauer, J.I. (ed.) 1997. Placing nature. Culture and landscape ecology. Island Press, Washington, DC.

Naveh, Z. 2000. What is holistic landscape ecology? A conceptual introduction. Landscape and Urban Planning 50:7–26.

Naveh, Z. 2007. Landscape ecology and sustainability. Landscape Ecology 22: 1437–1440.

Naveh, Z. and Lieberman, A.S. 1984. Landscape ecology. Theory and application. Springer-Verlag, New York.

Odum, E.P. 1989. Ecology and our endangered life-support systems. Sinauer Associates, Inc. Publishers. Sunderland, MA.

Odum, H.T. 1983. Systems ecology: An introduction. John Wiley & Sons, New York.

O'Neill, R.V., DeAngelis, D.L., Waide, J.B., and Allen, T.F.H. 1986. A hierarchical concept of ecosystems. Princeton University Press, Princeton, NJ.

Pearson, S.M. and Gardner, R.H. 1997. Neutral models: Useful tools for understanding landscape patterns. In: Bissonnette, J.A. (ed.), Wildlife and landscape ecology. Effects of pattern and scale. Springer-Verlag, New York, pp. 215–230.

Prigogine, I. 1993. Le leggi del caos. Lezioni italiane. Editori Laterza, Roma-Bari.

Pulliam, H.R. 1988. Sources-sinks, and population regulation. American Naturalist 132: 652–661.

Pulliam, H.R. 1996. Sources and sinks: Empirical evidence and population consequences. In: Rhodes, O.E., Chesser, R.K., and Smith, M.H. (eds.), Population dynamics in ecological space and time. The University of Chicago Press, Chicago.

Rapport, D., Costanza, R., Epstein, P.R., Gaudet, C., and Levins, R. 1998. Ecosystem health. Blackwell Science Inc., Malden, MA.

Real, L.A. 1993. Toward a cognitive ecology. Trends in Ecology & Evolution 8: 413–417.

Richtie, M.E. and Olff, H. 1999. Spatial scaling laws yield a synthetic theory of biodiversity. Nature 400: 557–560.

Turner, M.G. 1989. Landscape ecology: The effect of pattern on process. Annual Review of Ecology, Evolution and Systematics 20: 171–197.

Turner, M.G. 2005. Landscape ecology: What is the state. Annual Review of Ecology, Evolution and Systematics 36: 319–44.

Turner, M.G., Pearson, S.M., Romme, W.H., and Wallace, L.L. 1997. Landscape heterogeneity and ungulate dynamics: What spatial scales are important? In: Bissonnette, A. (ed.), Wildlife and landscape ecology. Effect of pattern and scale. Springer-Verlag, New York.

Ulanowicz, R.E. 1997. Ecology, the ascendent perspective. Columbia University Press, New York.

von Bertalanffy, L. 1969. General system theory. George Braziller, New York.

Whittaker, R.H. 1975. Communities and ecosystems. MacMillan Publishing Co, Inc., New York.

Whittaker, R.J. 1999. Scaling, energetics and diversity. Nature 401: 865–866.

Wiens, J.A. 1997. Metapopulation dynamics and landscape ecology. In: Hanski, I. and Gilpin, M.E. (eds.), Metapopulation biology. Ecology, genetics, and evolution. Academic Press, San Diego, pp. 43–62.

Wiens, J.A. 1999. Landscape ecology: Scaling from mechanisms to management. In: Farina, A. (ed.), Perspectives in ecology. Backhuys Publishers, Leiden, NL, pp. 13–24.

Wiens, J.A., Crawford, C.S., and Gosz, R. 1985. Boundary dynamics: A conceptual framework for studying landscape ecosystems. Oikos 45: 421–427.

Wu, J. and Hobbs, R. 2002. Key issues and research priorities in landscape ecology: An idiosyncratic synthesis. Landscape Ecology 7: 335–365.

Zonneveld, I.S. 1995. Land ecology. SPB Academic Publishing, Amsterdam.

Chapter 2
Toward the Essence of the Landscape: An Epistemological Perspective

Introduction

The landscape is a recently investigated domain of the real world. The number of studies with the landscape as a focal object has grown exponentially since the 1980s. Many definitions of the landscape and its theoretical and applied subdisciplines are present in the literature, often creating great confusion about its vocabulary and ideas.

To form a new science we need recognized principles and efficient tools; both ingredients are already in the basket of landscape principles. The principles come from different theories that show extraordinary convergent properties: the General System Theory (von Bertalanffy 1969), the Semiotic Theory (Eco 1975), the Autopoiesis Theory (Maturana and Varela 1980), and finally the Ecological Theory (Scheiner and Willig 2007).

Among the various sciences that have found the landscape to be a possible subject to be studied and managed, ecology has the privilege to approach this study in a holistic way (sensu Naveh 2000), mixing together different components of environmental complexity.

The studies that have been carried out in recent years have been based upon the conviction that the landscape is an entity rich in information that has emergent properties. These patterns are the product of the processes and vice versa, that most of the processes are scale-dependent and that a hierarchical organization forms the basis of the landscape, are other fundamental assertions of these recent studies.

Furthermore, there have been many disputes about the definition of the landscape and about the opportunity to consider the influence upon the landscape domain of human processes, for instance economic processes, and cultural and political issues (Naveh 2000).

For many landscape ecologists, the landscape is a large container of different processes that interact with each other to create the observed complexity. For others, this vision is metaphysical and not scientifically correct, so it is rejected. Finally, for others, the landscape is a geographic (physical) space in which many different phenomena that can be observed are formally described.

It is time to create order in this young discipline, and also to respect the expectations of developers, environmental managers, and policy makers.

A. Farina, *Ecology, Cognition and Landscape*, Landscape Series 11, DOI 10.1007/978-90-481-3138-9_2, © Springer Science+Business Media B.V. 2010

In the first chapter I presented landscape ecology as a relevant discipline tracing back its recent story of successes and failures. It is now time to discuss new ideas and concepts with the hope of bringing order to a field in which basic research and applications are often confused from an epistemological viewpoint. The recent criticisms made by ecologists largely depend on the use of a gestalt approach from practitioners and applied scientists on one side, and on the use of too many mathematical models by spatial ecologists on the other.

First of all, it is necessary to define the subject (landscape) and secondly the domain of such a subject which is not an easy task according to the premises and discussion in the first chapter.

The landscape can be considered contemporarily as the container of physical and cognitive entities (Fig. 2.1). Cognitive entities are expressed by rules and values; physical entities by soil-water categories, plant or animal spatial aggregations. Cognitive entities are represented by a network of eco-semiotic interactions between organisms and the species-perceived surroundings.

Fig. 2.1 Physical and cognitive properties characterize the landscape "entity" but for both the properties borders and delimitation are necessary

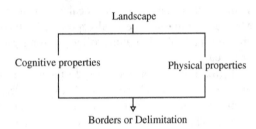

The Necessity of Defining a New Science of the Landscape

In the recent congress of the International Association of Landscape Ecology (IALE) in Darwin (2003), many empirical studies confirmed that the patterns and processes occurring in the landscape can capture the interest of science and of politicians and practitioners as well. This contribution presents a vision according to a theory of the landscape and the way studies of the landscape can be so different and peculiar as to impel us to create a new science of the landscape.

Science is the entire universe of theories, paradigms, principles, and operative tools that regard a specific aspect of our world. Or one can define a science as the emergent property of knowledge on a distinct field. To receive full achievement and validation, a science needs several cultural tools, starting from ontology, philosophy, epistemology, and semiotics. Mathematical models should be anticipated by conceptual and linguistic models.

Every science requires a specific language and landscape science is not an exception. For this reason, when we start to describe the landscape we are strongly encouraged to adopt a language that can describe exactly the phenomena with which we are dealing.

The landscape is an entity shared by different philosophies, different paradigms, different methods and scaling. It requires a common semantic basis and related principles.

The Landscape Domain

We will use the word "domain" to indicate the context in which we are operating, the universe in which a process occurs, evolves, or is maintained, a field in which some phenomena occur in an exclusive way.

Often we erroneously describe phenomena as being apparently linked to each other but in reality they belong to different domains and the landscape as an entity must be referred necessarily to a specific domain.

In other words they do not all occur together, but most phenomena occur in separate domains. Hierarchy theory is an attempt to describe these domains; unfortunately, this theory has created confusion because, as explained, it has hierarchical nested elements when in reality the described items are parts of different domains. In fact domains are not necessarily hierarchically arranged and often have a parallel reference. When many domains share common elements we can use the word "meta-domains."

Many meta-domains contribute to the definition of landscape: environment, culture, economy, religion, and policy are the major composing domains.

Domain is very close to the niche concept. The niche in ecology is the range of the entirety of biological characters. We can first make a distinction between non-corporeal (cognitive) and material (physical) domains. This distinction is important whether we consider ourselves as observers or as organisms directly involved in biological processes.

Human processes are referred to in five main domains: religion, culture, economy, policy, and environment. Apparently divided and separated they are in reality the meta-domain of humanity.

Religion, culture, economy, and policy pertain to a human meta-domain, and environment is apparently distinct. But most human actions have two main targets, humanity "per se" and the environment. Actions toward humanity try to modify uses, rules, ethnic domination, economic and health fitness. The actions toward the environment try to optimize the use of resources. Humanity is strongly influenced by decisions at every level of social organization, and it seems useless to discuss it. In modern societies, policy occupies a central place in this matter, protruding its tentacles in every direction of humanity. Policy is a very active process that rejuvenates the physiology of human societies. The mechanisms that self-guide the society are like fuel in a machine. The fuel is represented by new ideas about societal organization, the personal will to dominate the economic system. All are ingredients that are considered under the umbrella of policy. Morality and ethics rarely play a central role, and often it is only public agreement that guides the process, at least in democracies.

Conversely, religion is a conservative domain that tries to stabilize human processes at every scale and in every type of society. Religions create a non self-referential domain, conversely, culture, policy, and economy are self-referential domains that act through a circular closure and have an adapting behavior according to the feedback between singular components.

It is our opinion that landscape ecology should be considered as a science that studies the integration of these different domains and meta-domains. The influence of every domain is not considered in modern sciences because each discipline has been created in limited and distinct domains. Landscape science should consist of the product of convergent processes and integrated meta-domains.

The landscape domain is the highest level at which we can consider the complexity human beings perceive. This domain embraces many other domains and the elements that compose this domain are other distinct domains.

Physical and cognitive domains are the proximate domains to the landscape domain. Other meta-domains can be found at the periphery of this domain. In order to understand this book, it is necessary to accept this assumption. Humanity can live without the landscape paradigm, but a task of culture and particularly of science is to help common people to better understand the complexity of the living and educate them to tolerate and accept different perspectives. Finally, the tools to investigate such meta-domains are extracted from many disciplines including ecology, biology, sociology, economics, and psychology.

Three Phenomenological Domains of Landscape

We distinguish three phenomenological domains that characterize the landscape entity (Fig. 2.2):

Fig. 2.2 The three possibilities to approach the landscape paradigm: Neutrality-based landscape (a landscape exists in any way, although patterns and functions are not distinguished by the observer), Individual-based perceptional landscape (the landscape is the part perceived by a species), and finally the Individual-based cognitive landscape that is a particular status of intentional description of the surroundings (see Dennet 1983)

Neutrality-based landscape   Individual-based cognitive landscape

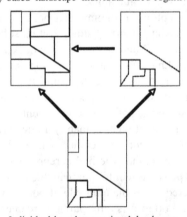

Individual-based perceptional landscape

(1) Individual-based perceived landscape. The landscape is the result of organism perception, and is a species-specific entity. In this case a specific landscape exists for every organism or process that selects from the neutral landscape the parts with which it enters into a perceptual relationship. This idea can be extended to processes, considering the "behavior" of the processes. It could seem a forced manipulation of the reality, but we discuss this point in detail later, trying to demonstrate the efficacy of the approach.

(2) Individual-based cognitive landscape. The landscape is the result of cognitive elaboration. In this case the landscape is considered as a system, and the observer opens a window on the system. Knowledge of the system is only partial, and is reduced to the spatio-temporal window of the observer. The observer interacts with the object observed and modifies its descriptive attributes, and the same observation becomes a perturbative entity to the observed object.

(3) Neutrality-based landscape. The landscape is a neutral entity that exists without organism interference and interpretation. In this case, the landscape is composed of functional and structural units created by the aggregation of individuals of the same species (populations) and different species (communities). The aggregations have self-organizing capacities under the pressure of stochastic events. A plethora of studies in landscape ecology have assumed this perspective. The neutrality-based landscape is composed of a mosaic of patches created by aggregation of individuals and by spatial processes. According to this perspective the landscape is a source of possible perceptions and/or cognitive interpretations that depends on the sensory perception of organisms. The neutrality-based landscape is a permanent source of information available for evolutionary processes. The importance of this perspective is underestimated in the adaptative and evolutionary processes of living systems.

In the chapter dedicated to the ontogenesis of the landscape (Chapter 5), I will discuss in more detail the role of the three perspectives. Every perspective contributes to the organization of a complete landscape that is the meta-domain created by the domains composing it.

In order to respect the paradigm of environmental complexity, we have to accept all three of the perspectives which will be discussed in detail in the following chapters.

It is frequent in the realm of science to work inside specific areas without considering other areas that could address the same problem in different ways. Unfortunately, we fail to consider several realities according to the different sciences. Rock and earth dynamics are described by the geological sciences, and we associate soil and earth with such problems. Finally we are convinced that geology is something that exists. The same story applies to biology: We separate the living organisms from their environmental context, and the discipline moves from a descriptive realm into an existential realm.

Such an attitude separates the world into compartments out of which it is extremely difficult to explore the inherent complexity. For this reason, the landscape is an existing reality for ecologists and geographers, but not for economists or social scientists.

The "real world" is the product of description of a reality by specific disciplines, which produces a dramatic separation of this description into unique worlds that have difficulty communicating a complete understanding of the planet's complexity. In the end we are convinced that geology, ecology, and geography are separate components of the complexity, even though they exist only in our mind!

Why do we now ask for a new science? The reply to this question is complicated. Today humanity has a global view of its problems. At any time today we can be aware of events that happen in different parts of the world. Events are the basic component of our dynamics. The observer (individual, society) has a planet-scaled vision of the problems and is aware of direct and indirect influence of processes that are separated only at their origin. Like the water in a river catchment, separate springs convey water in the main streams and channels until they coalesce into a single river.

It must be considered in this metaphor that landscape science is the main river. If you move in to examine the details, or the individual headwater springs, you can distinguish the separate domains. The hierarchy and the relationships between domains are responsible for the observed complexity.

When we adopt a landscape view, it is likely to be at the mouth of the river. It is not possible to distinguish the origin of the water. The different typologies of water have been blended together, and are now an undistinguishable collection. Often we analyze the details, and from their sum we reconstruct the complex system produced. This procedure is inconvenient and inefficient in terms of advancement of human thinking. The procedure should be reversed so that we start from the complex systems, and, by their emergent properties, we can move toward the "spring" of the catchment and discover its details.

Complexity and Domains

The world is complex and its complexity can be explained not by compressing the components of a system, but by distinguishing the complexity as composed of different domains that operate independently and at the same time influence each other. Every domain operates in its space-time window and has properties that are conditioning entities and processes in that specific domain.

In many cases the domains can operate in strict relationship, and in this case we can consider constellations of related domains like meta-domains.

A domain is a context in which specific processes occur and the phenomenon that exists in a domain is domain-specific. A domain is not an extra-sensorial entity but a physical, perceptive, and cognitive space composed of entities that interact for a specific process that exists only in that domain. A domain can be considered in ecology as the niche of a species, the range in which certain variables exist and operate.

A watershed, the area in which atmospheric, surficial and underground water move toward a coalescence sink is a geographic domain. A watershed is clearly independent of another neighboring watershed but a net of watersheds can create a large river catchment. In this case every distinct watershed can be considered as a part of a catchment meta-domain. Similarly a political domain creates an independent country with distinct laws, uses, and ethnic populations that operate inside a specific area.

Is the Landscape a Unit or a System?

For many scientists, a landscape is a large area in which it is possible to study processes and patterns that have huge dimensions, but landscapes are a matter of scale; are they then entities/units or simply a domain?

If we observe large areas, for instance from an airplane, is this area actually a landscape or is the landscape a cognitive domain?

The major problem in defining the landscape consists in the duality by which we can see the landscape as a unit or as a system.

If we consider the landscape as a unit, we have to recognize the character of such a unit, specifically the autopoietic character, or, in other words, the capacity for self-organizing and self-maintaining through the use of inner forces and processes. If the landscape is a unit, it must have a defined border (Fig. 2.3) and must be distinguished from its surroundings.

A landscape as a unit would have the character of an organism and would behave as an autopoietic entity, with a circular closure dependent on inner characters and weakly influenced by external input.

The problem in defining a unit (conceptual or physical) is to distinguish it from its background. And one needs a large-scale vision to include the unit on a background. Often we select for our convenience a "piece of land" and we define this

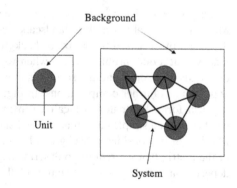

Fig. 2.3 Landscape can be defined as a unit when it works in isolation like an island in the ocean. All the properties are inside the unity. A landscape is a system when the combination of separate entities creates emergent properties such as connectivity. A system is based on processes

as a landscape unit but this operation is without sense and is destined to fail. If we consider the landscape as a system, it exists only if the composing parts interact with each other.

For instance, a man is a unit, a separate autopoietic entity, but contemporarily is a part of the society. In this latter case, he is part of a society if there are other men around, but if he is Robinson Crusoe he is only an entity.

A landscape is at the same time a unit and a system, depending on the role and the local conditions. As a consequence of this fundamental distinction, the approach used to investigate properties and the related metrics must be selected consequently.

Unit and system views have in common the space. This is a very important element of distinction of a landscape from other ecological agencies that we will discuss later.

Assessing the Characteristics of the Landscape: Ecosystem Versus Landscape

In the past decades many ecologists have used ecosystem and landscape synonymously. It is time to make a clear differentiation between these two mental representations of reality.

Ecosystem is a term that can be applied to all (spaceless) mechanisms that operate in nature, and can be considered the functional framework.

Functions and relationships, flow of energy and material, food chains are some of the distinctive processes connected with the ecosystem concept.

The landscape framework couples functioning with the space in which such functioning operates. As a consequence, the landscape framework is inclusive of the ecosystem framework. Often in literature a landscape is considered a system of ecosystems, but this is not completely correct. The ecosystem framework pertains to a different descriptive domain of the landscape domain.

The Landscape as a Matrix

A very popular vision of landscape is a matrix of processes and related patterns. Most of the actual research in landscape ecology focuses on this perspective. The described objects in this matrix, or background, have their domain inside such a matrix. But a dual problem arises when we follow this vision. The first is how to distinguish a matrix, and the second is how to distinguish the patches inside. This approach has further complications. In Fig. 2.4 the problem is represented in geometrical form. If a patch is localized inside a matrix, the relationship with other patches is not a simple case of absolute distance but a matter of matrix localization. Matrix can be considered background, but, according to the delineation of a part of the matrix that we require to describe any phenomenon, the position of patches depends on the size and localization of the matrix window. Matrix concepts are

Fig. 2.4 According to the position of the describing window a matrix can describe different objects (A,B) in the same domain

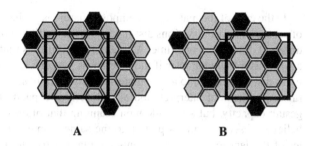

A   B

implicitly undetermined although we believe the contrary. A possible way to bypass this uncertainty is to consider a domain and assume contemporarily that such a domain has heterogeneous characteristics, but still the choice of the observing window is not determined by objective characters. In modern landscape ecology, we assume that the landscape is a heterogeneous entity and that in such an entity we can observe a patch that is distinct from the background. According to this view, it would be possible to distinguish one patch at a time and not all the patches of a matrix, because in this case the matrix would be coincident with the background and, as a consequence, invisible. The matrix concept can be considered a first approximation that is not the product of observations but the result of an anticipated vision of the environmental context. It is a matter of sequential scanning of the background and the assemblage a posteriori of the different units detected one at a time.

When we try to attribute a function to a matrix, we assume a different role for the neighboring patches as though they were at the same time real. In effect every patch that we can distinguish pertains to different domains, or if you prefer every patch is the result of a distinct process (Fig. 2.5). That they are neighbors is a matter of observation, but they are false neighbors. Maintaining this assumption we are obliged to reject most of the interpretations that we have accumulated especially in landscape ecology.

The next step is to recognize a composite unity, a unity in which it is possible to distinguish different components. In this case the heterogeneous nature of such unity develops solitary characteristics and thus enters into the domain of the unit.

Fig. 2.5 Most of the processes that we can observe pertain to different domains (*a, b, c, d, e, f*) but are perceived by our sensors as contemporary and topologically related. This vision often is responsible for the bias that we introduce in the management of natural and human resources

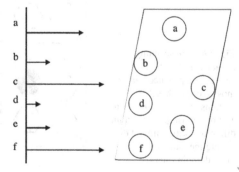

In the case of the matrix, the composite nature is probably not real, but is an effect of accumulated observations distinguishing each unit at a time, and we are exploring different domains at one time. A residual possibility is that a matrix defined in such a way has a true character if the patchiness pertains to a meta-domain. Our capacity to distinguish and to accumulate such distinction, such as in a scanning process, raises problems of interpretation of reality. This capacity must not be considered a gestalt capacity, but a multidomain scanning that often swindles the observer who believes observed entities pertain to the same domain. The heterogeneous character of the landscape that represents the major attribute used in landscape ecology to process complexity could be a matter of human capacity to scan unities from a background using a multiresolution sensor. This hypothesis is extremely unpopular at present because scientists do not pay attention to the reality but seek to find evidences to a "pre-ordinato" mental model. We assume the heterogeneous character of the landscape, and we don't pay attention to other possibilities. In Fig. 2.6 the procedure that we use to scan a matrix is represented.

The problematic linked to the matrix bias affects most of our information on the real world and produces a great epistemological confusion. We discuss this later in the mosaic theory and for the moment we try to clarify the relationships between the different domains, a matter that we solve by adopting the ecotone paradigm in landscape ecology.

We have to assume that heterogeneity is represented by an asynchronous perception of separate unities recombined in our mind. Several problems in interpreting the reality of our surroundings are due to this distorted vision of reality.

A duality of problems must be discussed now that, for the first time, we have rejected the assumption of the universal value of heterogeneity. However, the heterogeneity does not exist at all; it is a product of the observer. We have to distinguish between a simple entity and a composite entity. The first is composed of a simple structure. It could be a strand of thread of a unique color and material (like wool). A composite entity is composed of several parts that are associated into the functional whole.

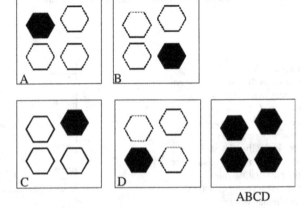

Fig. 2.6 The scanning capacity of the observer agency can produce a false mosaic of patches that pertains to different domains. The relationship between the different entities are not direct but are mediated by meta-domain filters

When woodland is in contact with an agricultural field, the two entities interact with each other in a special location that we call an ecotone. This is correct from a formal point of view and especially if we adopt the paradigm of geographical heterogeneity. But if we observe the processes that create and maintain these two entities it is clear that the processes responsible for a woodland growth are quite different from the ones that produce a crop field. The exchange of information between the two systems is quite low; order is produced inside the separate systems and not between. The domain in which a forest grows and develops is different from the domain of the human being. At this point we have to reconsider carefully the ecotone paradigm that is not a simple juxtaposition of patches but areas of contact between different systems or domains.

In regard to the typology of different ecotones, it is necessary to introduce a differentiation between intradomain and interdomain ecotones. In the first case the flux of material and energy might be higher than in the second.

For example, such intra and interdomain ecotones can be represented in moving outward from the center of a city along a gradient of "naturalness."

Space and Landscape

Landscape is defined in a physical realm. This is quite clear and not considered an element for further discussion. But if we select a space in which we say, "this is my landscape of interest," are the borders real borders or are they borders that cope with some selected processes?

Two main views can be discussed now. The first is that a landscape is composed of functional units, independent from our observation, and the second is that these units are attributed to a space.

We strongly believe that both systems exist. In the first case the delineation of the space is not produced by the observer, but by an observer-independent process. The observer recognizes the place in which a physical discontinuity appears, like between savanna and tropical forest, or montane treeline. In this case it is not a matter of scaled observation.

This explanation seems very simple and direct, but is strongly conditioned by the limits of discrimination that are inherent to the observer. Are there other limits that are outside our range of observation and that are responsible for processes later detected?

Finally, we can address two different concepts: the visible (coupled) and the invisible (un-coupled) landscape.

The visible landscape is a system that is delineated by tracking limits observed directly and by a coincident vision of the processes that are responsible for such limits: Processes and patterns observed at the same time, are coupled.

For instance, when one observes the shadow under a tree, one can distinguish the shape of the tree and one can link this form with the position of the sun light and thereby fully understand the process of shading.

But when one observes the colonization of trees in an abandoned field, one's vision is limited to the patterns created by the vegetation but one can not observe the process that has generated such modification. The abandonment of the field pertains to a socio-economic domain not visible using physical sensors. The land's abandonment can't be visualized, but the effects can be visualized. It should be an error of belief to explain just in terms of secondary succession. This method of self-explanation is superficial and not correct from an epistemological point of view.

Maturana and Varela (1980) have defined space as "the domain of all the possible relations and interactions of a collection of elements that the properties of these elements define." It is not difficult to adopt this idea of space as synonymous with landscape, as is intended today by most authors. This vision may appear too general, but it is exactly the way in which our science and society address the problems, and from which springs the concept of complexity. Time requires new concepts like evolution and adaptation, processes that require successive steps. Landscape is a domain in which time allows different entities to move into organizational status and dynamics. Time is a necessary ingredient to create an organized entity: it is a process of organization, or, is the organization per se. Organization refers to a sequence of acts that move an entity from one space to another, assuming functions. Time and organization can be considered two faces of the same phenomenology. But since you can't destroy the time, similarly, you can't destroy organization. You can modify organization, substituting with another form of organization. Even a paroxysmal event does not destroy. Rather it modifies this organization, substituting elements with others and respecting the memory of the system.

Memory and Landscape

"What is memory?" is not a simple question. In neurophysiology, memory is the storage of past information that can be retrieved when a signal is sent to the memory blocks. Memory is fundamental for our life because it links together processes and allows one to use past experience in a new condition. Memory to our brain is like soil to plants.

The memory of the landscape resides in our culture and is accumulated in the physical world. For instance, soil is the recent memory of bio-physical processes. A deep soil has a more long-term memory, and sedimentary rocks have a very old memory. "Fragments" of atmosphere can be detected inside the Antarctic ice pack, representing a memory of the time a thousand years ago when the ice pack first formed.

Memory accumulates events everywhere. Only a new process at an infinitesimal unit of time has no memory. The accumulation of events and materials shapes the landscape. The landscape has a memory accumulated in physical as well as cultural domains. Often some types of memory are hidden and cultural mechanisms are necessary to rediscover and interpret the accumulated information. The terracing in many mountain regions is an example. The patterns of terracing cannot be immediately explained, but rather it requires the knowledge of past land use. At the same time a charcoal plaza across beech forests of the northern Apennines is not recognized by tourists that are in transit. The pollarding of the past can be observed today as a stand of strange trees; in reality the tree shapes observed today are the result of past use of these trees.

A landscape is like a roof composed of tiles. Every tile is posed on the margin of a precedent tile. Often the evolution of the culture decouples patterns from processes, exposing the landscape to an apparent novelty. Like tiles allow water to flow, at the same time the landscape tiles allow the landscape processes to move forward. Memory is the accumulation of remnants of past processes not included in the reorganization of present-day processes. This distinction is not easily to be proposed, because there is a duality in the memory: first, a new process starts using as its basis an old one (implicit memory like a universal value): secondly, the new processes can use only a part of the past elements that persist or better survive to the new one (witness memory). Of course the importance of the first memory is larger than the second, but nevertheless the witness memory often is very important for our cultural memory too! Humanity needs to maintain a relationship with the past and our distinction of time in past, present and future is strategic for our actions and for our mental health.

The present-day landscape is inside a temporal domain that is contained in an asymmetric window, with the larger part open to the past (memory) and a narrow window projected into the future (Fig. 2.7).

Memory is not a place in which to draw up elements (processes, entities) but an organizational status that influences further successive status. The irreversibility of organization is easily demonstrated. An old tree can't be reduced to a young tree! Apparently a forest can be reduced by fragmentation to small portions of isolated trees as in the early successional stages, but this is not a primary succession. It is a fragmentation process, very different from the secondary succession, although the resulting patterns are very similar. Organizing procedures involve the investment of free energy that is captured inside a structure. When organization is dismantled, energy is released by the change in entropy. The balance between the two opposite and contrasting trends is obtained by the use of energy.

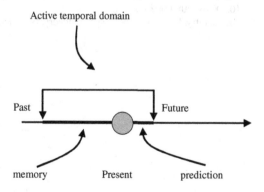

Fig. 2.7 Processes of the present day need a tail from the past and a head in the future. This requires memory and prediction that delimits a temporal window

New Perspectives

When we consider the landscape, we have several possibilities to address this topic. But is the landscape an entity or part of a system? Or is it simply the component of our perception, with the perception entering into the domain of the biological phenomena, and definitively an agency? We start considering the three possibilities as true, and during my discussion we will try to clarify my position and also the positions that today are considered applicable to the scientific method.

The landscape perspective is one of the most relevant and complete possibilities to understand the complexity in regard to we are not only intruders but also observers and actors. We must move from a static vision of the world subdivided into rigid compartments that are the object of the modern disciplines to a world in which there is totality (gestalt) that imparts order to every system as a result of the action of several living actors.

Suggested Reading

Allen, T.F.H. and Hoekstra, T.W. 1992. Toward a unified ecology. Columbia University Press, New York.
Cillier, P. 1998. Complexity & postmodernism. Routledge, London.
Maurer, B.A. 1999. Untangling ecological complexity. University of Chicago Press, Chicago.
Merry, U. 1995. Coping with uncertainty. Praeger, Westport, CT.
Waldrop, M.M. 1992. Complexity. Simon & Schuster, New York.

References

Dennet, D.C. 1983. Intentional systems in cognitive ethology: The "Panglossian paradigm" defended. Behavioural and Brain Sciences 6: 343–390.
Eco, U. 1975. Trattato di semiotica generale. Bompiani, Milano.
Maturana, H.R. and Varela, J.F. 1980. Autopoiesis and cognition. The realization of the living. Rediel Publishing Company, Dordrecht, Holland.
Naveh, Z. 2000. What is holistic landscape ecology? A conceptual introduction. Landscape and Urban Planning 50: 7–26.
Scheiner, S.M. and Willig, M.R. 2007. A general theory of ecology. Theoretical Ecology 10.1007/s12080-007-0002-0.
Von Bertalanffy, L. 1969. General system theory. Braziller, New York.

Chapter 3
Toward a Theory of the Mosaic

Introduction

Whether viewing our planet from space or simply viewing the structure of soil we observe objects spatially arranged as mosaics. The mosaic seems to be the common pattern that we perceive, especially from an aerial view.

We can observe mosaics everywhere around us moving from the centimeter-sized scale of lichens to the megascale (thousands of kilometers) of biomes. Cloud systems, frozen water, cultivated fields, the distribution of plant communities, and animal flocks are either living in mosaics or are organized in mosaic-like structures. Origin, function, and evolution of the observed mosaics can be very different, and, as in evolution of the morphologies in plants and animals, we can observe convergent patterns in the spatial arrangement of living and nonliving objects in water, on the surface of soil, and in the atmosphere.

In this chapter, I will try to determine whether this is a general pattern probably generated by common processes or an evolutionary trend in the direction of the mosaic. Commonalities of such processes are probably connected with rules that until now were not explicitly known or that have been explained only from a physical point of view.

If life is cognition, cognition has emergent properties that are associated with mosaic-like patterns, and this is not a simple matter of scale but a matter of interacting phenomenological domains.

We discussed in Chapter 2 the definition of the unit that must be distinguished from the background, and we argued for the concept of landscape as a functioning entity. The landscape exists in an organized space, and organization means time, information, and probabilistic status.

In this chapter we will try to describe the dominant patterns observed in a geographical space, while aware of the difficulties of introducing a bias between the descriptive domain and the operational domain. The heterogeneity, such as the observed patchiness of a landscape, is a reality.

In this chapter, we address questions such as:

Why are organisms organized in space in a mosaic?

What are the consequences of the mosaics on the living functions of the
organisms?

How is the theory of landscape science consistent with the mosaic perspective?

Ecological Complexity and Mosaics

Ecological complexity is the result of several processes that are determined by at
least four main properties that behave differently: invariance, density, diversity, and
spatial arrangement (Fig. 3.1).

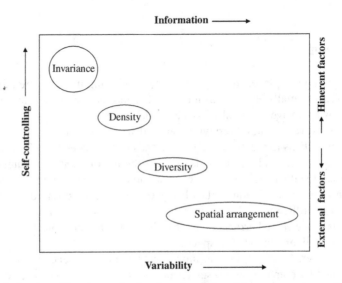

Fig. 3.1 Ordination of invariance, density, diversity, and spatial arrangement according to self-
control and variability

Invariance means the capacity of individuals to maintain constant forms from one
generation to another and at the same time the capacity to adapt to a changing world.
This factor is largely controlled by genetic mechanisms acting in an evolutionary
fashion.

Density represents the quantity of individuals necessary to maintain populations under the numerical fluctuations by natural stressors and environmental
unpredictability. Density means populations and their emergent characteristics that
regulate the number of individuals and their behavior.

Diversity assures the coalescence of different species into communities with beneficial effects on single species by multiple uses of available resources. Diversity
allows the optimal use of resources by the creation of complex trophic linkages,
interspecific competition avoidance, and mutual reinforcement.

Finally, spatial arrangement provides the ability to cope with the environmental variability in the physical space. The different arrangement of patches in a
mosaic greatly contributes to the creation of new conditions for species, populations, and communities. The spatial arrangement of environmental patches presents

the highest level of spatial arrays, which contributes in great measure to "creating" new possibilities for communities and populations. The complexity of ecological systems is created by the processes activated by the above-mentioned factors.

Every factor is sensitive to the environmental constraints. The self-control of each of these factors changes greatly. Invariance is controlled by the same individual through distinct genetically driven mechanisms. Moving from populations to spatial arrangement, the internal control decreases in terms of importance and the external factors become more and more important. Finally, the spatial arrangement of patches becomes the level at which human activity manipulates, but only indirectly, diversity, populations, and genomes.

Variability increases as it moves from the first factor (invariance) to the last (the spatial arrangement) and this emergent property plays a fundamental role in assuring new opportunities for species adaptation. We recognize in the spatial arrangement of the objects (communities, populations, individuals) one of the most important sources of variability and unpredictability that largely contributes to self-maintenance of the ecological complexity.

If the mosaic is a universally recognized pattern of living and nonliving entities, a common ontogenesis should be recognized. If a horse is introduced to a new paddock and we observe the modification of the paddock through grazing and trampling (disturbances strictly associated with the horse), this introduction gradually produces more and more disturbed areas connected by pathways. Finally, after a few days the horse will have transformed the paddock, initially homogeneous in terms of grass cover, into a mosaic of differently grazed patches, connected by trails. What lesson can we learn from this? A mosaic is not per se generated like the pattern is observed, but by a differentiated intensity in local disturbance associated with a strong investment in memory. The reaction of the grass cover is not manipulated by an explicit goal-function of the horse but by the frequency of return of the disturbance and the direction from which the disturbance is coming. The system apparently is driven by external factors (e.g. grass species preferred by a horse), but progressively the external factors are associated with an internal dynamic of grass cover that reacts to disturbance, favors species more resistant to trampling, and generates a less diverse grass cover with the appearance of more homogeneous patches: the mosaic appears and progressively these mechanisms are reinforced! The disturbance intensity produces a mosaic that can survive, when created, for a long time by an intrinsic reinforcement.

Definition of Mosaic

The word "mosaic" is from the medieval Latin "*musaicus*" and means a creation from the Muses (Nine Greek divinities of arts and sciences, daughters of Zeus and Mnemosine). Mosaic was originally used to describe the artistic creation of figures on pavements or walls using small pieces of painted material, such as stones, marble, wood of regular shape (from the Greek *tesserà* (gonos); in Latin *tessera* means an object with four sides) placed close to each other to build figures (Fig. 3.2). Later the

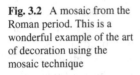

Fig. 3.2 A mosaic from the Roman period. This is a wonderful example of the art of decoration using the mosaic technique

word mosaic came to be used in many contexts to indicate a pattern characterized by distinct elements aggregated to produce a system.

In geographical or ecological meaning the ecological mosaic can be defined as an aggregation of patches of different types in which the interactions are determined by the functions that this mosaic is developing. In other words, a mosaic is the representation of the emergent properties of component patches. When we are dealing with a mosaic the study or the management of a spatially heterogeneous unit in which heterogeneity is produced by size, shape, and distribution of composing patches is necessary.

This unitarian vision of a mosaic allows the investigation of phenomena that occur at a scale many times larger than the patch scale.

Enormous efforts have been undertaken by landscape ecologists to understand the role of mosaics, especially the one created by the combination of human use and vegetation dynamics; but this analysis has appeared often too superficial and too restricted to a relatively narrow range of temporal and spatial scales. Finally, patterns are considered in terms of spatial objects that maintain their properties independent of the context in which they exist. We have very scant information on the rule and functions of the mosaics, but probably the heterogeneous distribution of all the objects in our living and nonliving space pertains to a general rule of the matter. Nevertheless, the study of mosaics appears a promising approach to a better understanding of the complexity of our living system.

Dimension, shape, and contagion of the ecological mosaics are the product of interaction between organisms and their physical and biological context. If mosaics are common features in any biological and nonbiological world it is important to discover the rules and the processes by which they are created and maintained.

Heterogeneity and Mosaic

Heterogeneity is a common attribute of every ecological system and a lot of scientific attention has been paid to this theme (Kolasa and Pickett 1991, Hutchings et al. 2000). Species are living in environments that are highly heterogeneous in space and time. The heterogeneity is created by the variation in abiotic factors (physical and chemical properties of the soil, microtopography, microclimate) as well as by organisms (plants via leaf deposit, exudate, root growth; animals by grazing, trampling, burrowing). Some organisms, like termites, are real ecosystem engineers promoting heterogeneity and providing in such a way habitats for many other species that have a less direct relationship with the environment. Finally, another cause of heterogeneity is represented by stochastic events like earthquakes, hurricanes, fires, volcanic eruptions, flooding, etc.

Heterogeneity changes in time and space and according to age every organism may experience a different type of heterogeneity. For instance, a poplar seed experiences a different type of heterogeneity compared with an adult plant. A dramatic difference appears when we compare the heterogeneity perceived by a less mobile species like a caterpillar to that of a very mobile butterfly.

Heterogeneity is not a novel concept in ecology; nevertheless, the interest of ecologists is growing, thanks to the use of new technologies to explore this relevant pattern of ecological complexity.

The nature of heterogeneity is not as clearly evident as we believe. A heterogeneous medium means that the composing elements are different or have different spatial characters. Heterogeneity has been distinguished by Kolasa and Rollo (1991) into two types: measured heterogeneity and functional heterogeneity. The measured heterogeneity is "a product of the observer's arbitrary perspective" (Kolasa and Rollo 1991). The functional heterogeneity is the heterogeneity perceived by the ecological entity (organism or process).

According to a general perspective we can also describe two further types of heterogeneity: spatial and temporal. Spatial heterogeneity means a different distribution in space of ecological entities. Temporal heterogeneity means the change in characteristics for the same point across a period of time. Recently, Wiens (2000) recognized four forms of heterogeneity: spatial variance, patterned variance, compositional variance, and locational variance. As we will see later, the locational variance represents de facto the mosaic (Fig. 3.3).

Spatial variance, therefore, intercepts the irregular position of objects in the matrix, and patterned variance considers the importance of the nearest elements from a focal object. The compositional variance is expressed either by quality or quantity of the objects. Finally, the locational variance is the explicit spatial arrangement of patches in a matrix.

Heterogeneity and patchiness in some cases have been considered as synonymous. Heterogeneity normally is used to indicate the nonhomogeneous nature of

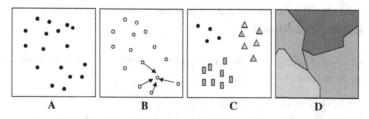

Fig. 3.3 The four main types of heterogeneity: (**a**) Spatial variance, (**b**) Patterned variance, (**c**) Compositional variance, (**d**) Locational variance (from Wiens 2000)

the environment in which a species is living. In other words heterogeneity is the grain of the habitat, the minimum dimension of objects perceived as distinct by an observer.

In general, a mosaic is a heterogeneous system. So in conclusion heterogeneity is a more general definition of variant attributes of an ecological system, and a mosaic is an entity pertaining to our cognitive perception.

Heterogeneity *versus* Mosaic and Disorder *versus* Order

Heterogeneity may be considered as a nonscalar image (far from the focus) of the environment, and when patterns emerge from heterogeneity, that emergence represents the focal point for that perception. In effect, heterogeneity per se is the status of the medium in which a species occurs, but we need to find the scaled dimension to perceive patterns. This means we must find patches and organization in a selected domain. In Fig. 3.4 a subset from a mountain pasture has been classified by using an automatic classifier. When the image is classified using 200 categories, the heterogeneity emerges. As we decrease the number of categories progressively, different patterns appear and a mosaic is represented (shrubs and fields). As stressed also by Wiens (2000), heterogeneity per se is not sufficient to describe a pattern; it is necessary to relate the heterogeneity with organisms.

Heterogeneity is a property of every matrix (medium), but without a scaling action such heterogeneity is simply a noise.

The human perception of heterogeneity can be associated with a disordered way to perceive the surroundings. Most of our surrounding is perceived as structures to which we can not attribute a precise function, and as a consequence these structures are objects without meaning that contribute to the background noise. The mosaic can be the result of disorder processes like fragmentation or conversely the result of a new order imparted to the system from outside drivers (Fig. 3.5).

Mosaic creation can be observed in different stages of evolution of a system: when a homogeneous system is heavily disturbed, like during fragmentation (sensu lato), and when a system evolves toward a more ordered system after a long undisturbed period. Shapes of these two types of mosaic could be very similar but the ecological processes are completely different. In the first case, we can call this mosaic far from order and, in the second case close to order.

Rules Governing Structure and Dynamics of the Ecological Mosaics

The distribution of energy, matter, and organisms on the environmental sub-stratum can appear in at least two different forms:

a – gradient-like, in a fuzzy system (Fig. 3.6)
b – border-like, distinguishing separate discrete units (Fig. 3.7).

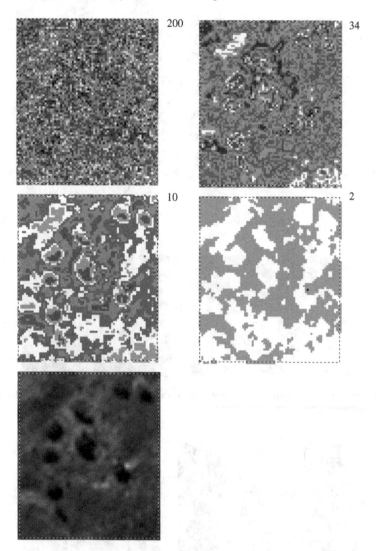

Fig. 3.4 The resolution at which a mosaic is observed can greatly change its appearance; for instance if we select 200 classes the image is indistinct, and if we select two classes the information is too compressed. Thirty-four classes is too much again but ten classes allows us to observe objects (shrubs) like in the b/w image

a: The gradient distribution pertains to many physical phenomena, such as the air and water temperature, sea water salinity, water content in the soil, and noise levels around an urban environment. Gradient means that particles are distributed according to a concentration threshold and that particles have the capacity to migrate following a dispersion behavior.

b: The sharp border between an entity and the neighboring one creates a mosaic, a puzzle of objects with different shapes, internal structures, and cognitive

Fig. 3.5 The mosaic may be created moving from order to disorder and vice versa, but the functions of these two types are completely different

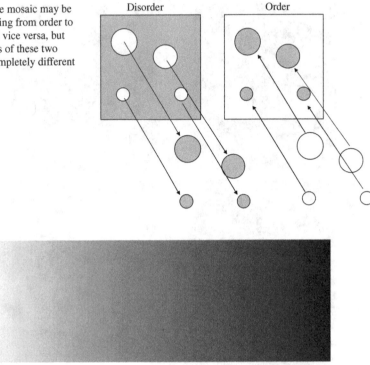

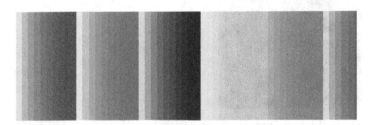

Fig. 3.6 Example of distribution of an event along a gradient, for instance the thermic behavior of an exposed soil

Fig. 3.7 In this figure we can observe a mosaic of large bands (6) and inside each band we can observe a gradient of ten tones

characters. The sharp border means that two different objects do not have the capacity to prevail one over the other, at least in the adjacencies, and that their distinct characters are homogeneous when observed. An example is from the patchy distribution of clonal plants like *Prunus spinosa* or nettles.

These two designs (gradient and sharp border), when observed at a different scale, modify their aspect and can interchange such characters. It is evident that a mosaic appears beyond the limit of a gradient, and that a gradient may appear when we reduce the scale of observation inside a patch of the mosaic, or when we enlarge the scale of observation.

In this case, the geographical distribution of *Prunus spinosa* is based on a gradient when locally it has a patchy distribution. And the gradient distribution of temperature shows patchy mechanisms when observed at the scale of climatic regions. Some conclude that, gradients and mosaics are a matter of scale. I think rather that they are two different conditions in which energy, matter, and life forms can be observed. The behavior of entities like temperature, water, and noise changes according to the level of scale at which we observe; different mechanisms are involved according to the spatial or temporal scale. How and when the different patterns (patchiness or gradient) shift is an openly disputed argument and, in my opinion, the key to understanding relevant parts of ecological complexity.

In general, it is reasonable to hypothesize that gradient patterns appear when a physical or biological entity is under the first stage of organization or when the dynamics are so strong so as to reduce the inherent behavior of a system.

The elements responsible for patchiness are local uniqueness, phase difference, and dispersal (Levin 1976). The mosaic appears when the system is under self-organizing control and the entropic state is moving toward a mature state. For instance, after heavy rains a stream can receive enough water to move in a chaotic (s.l.) way stones, gravel, sand, and particulate consistently. The more the flush is intense and short in time – for instance after a summer thunderstorm – the more the material is accumulated in a highly disordered way.

A mosaic appears when disturbance and recovery processes act contemporarily at different spatial and temporal scales independently. Observing the movement of stream deposits along the entire water body we can recognize zones with different dimensions of detritus. Moving from the head to the mouth the dimension of sediments change creating a "linear" mosaic of elements (Fig. 3.8).

A further example can be used. For instance, a recent seeded field with different types of grasses shows a mixed distribution of plants emerging from the random distribution of seeds during the seeding. In one or two years it is quite common to observe the appearance of a mosaic in which patches of dominant species are growing. This means that the system, in this case the plant community, moves to

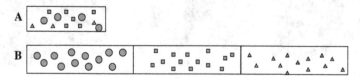

Fig. 3.8 In (**A**) the distribution of objects is random, a mixing of different objects is assembled by chance like during a short flush of a montane stream. In (**B**) a sequence of events, like annual floods distributes eroded material of a stream bed according to dimension and shape creating a linear, or gradient mosaic. In the first case time was the limiting factor in creating order in the system. In the second example order is much improved and the system shows self-organization

a coalescence in which the competition between plants, nutrients available, and micro-climatic constraints imposes order on the system.

What Processes Create Mosaics?

We have already discussed the origin and communality of mosaics. In general, a mosaic is created by the limited capacity of abiotic or biotic actors to spread everywhere. If we consider that every biological organism has a limited neighboring area in which to live and in which to influence the surroundings, we immediately understand that a mosaic is the result of such intrinsic character.

Every species lives in a restricted space and this produces "local" heterogeneity. If we add the tendency for every species to enter into contact with individuals of the same species, the clump effect is per se a mosaic. We can give an example by discussing phytophagous larvae such as caterpillars. Such larvae develop from eggs laid on the bottom side of a leaf and spread onto the upper part, biting the leaf surface according to a design that is strictly linked with the growing size of their body.

In this way it is possible to observe many irregular holes in the leaves that can be interpreted as a foraging mosaic. Every patch, which is in reality a hole in the leaf, is the result of destructive interaction with the host plant and the presence in the leaves of less-palatable parts, such as the vascular and sustaining system. Such a mosaic that has an irreversible effect on the attacked leaf is very similar to the one created by the macro-herbivore in grasslands (Fig. 3.9).

The mosaic is spatially distributed, with patches of irregular shape and size that reflect different behavior, different growing stage of larvae and also different climatic conditions (cold, hot, dry, wet, windy, etc.) that has favored or discouraged larvae to forage. The pattern created by these larvae is (apparently) easy to analyze because only one actor is involved. Nevertheless, the choice of the part of the leaf and the spatial arrangement of each foraging patch are not clearly understood by

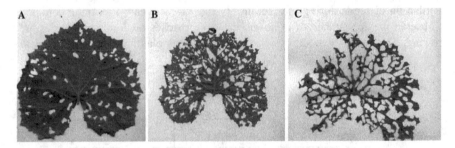

Fig. 3.9 Holes created by the feeding habits of caterpillar larvae can be considered patches in a system in which the vascular and skeletal system represents the landscape constraint around which animals try to optimize their grazing activity (leaf from a Beech forest, Northern Apennines). From (**A**) to (**C**) different levels of patch density show the increase of the "landscape" constraint in vascular and sustaining systems

just observing the behavior of larvae. And also in such a simple system based on the destruction of the resource a complex mechanism should act. Under conditions of abundance of resources, few holes appear and the distribution of foraging patches could be random, but, moving toward the reduction of resources, the location of foraging patches is more and more deterministic and finally is the main vascular and sustaining systems that form a constrained mosaic.

The same story can be observed when you are picking grapes in a vineyard. You move from one bunch to another after you have picked a grape. We can explain this in terms of instinctive anxiety under predatory pressure, but whatever is the cause of such behavior at the end you move from one patch to another, and you create unintentionally a mosaic of partially harvested bunches.

The Matrix as the "Container" of the Ecological Complexity

The ecosystem is defined as a complex level of organization in which animals, plants, and physical components interact. But it can also be considered as the context in which energy, carbon, or nutrient cycles operate.

The original (Tansley 1935) and the modern views of the ecosystem (Golley 1993) relate to the concept of function.

Although such conceptualization has been accepted worldwide, some aspects remain unclear. For instance, how can an ecosystem be delimited, how can we distinguish different ecosystems, and how can an ecosystem interact with another ecosystem? What are the relationships between other synthetic visions of our environment, like communities, meta-communities, and landscapes?

Some people consider a landscape as a system of ecosystems, but again this simplification of the reality is based on two assumptions (system, ecosystems) that, instead of reinforcing each other, create an exponential vagueness.

For this and many other reasons we try to view the complexity of interacting organisms, processes and related patterns in a very destructured way, calling a piece of real world simply an ecological matrix. Such a matrix is the reference for all the processes linked to physical and biological events.

The matrix in landscape ecology is defined as the major cover in which objects are interspersed. In a lightly fragmented forest, the matrix is represented by forests, and patches may be clearings or developed areas. This vision is useful to develop new ideas, but it is too geographically and descriptively oriented. In fact, when we compare the relationship between organisms and such mosaics, very often the searched for relationships appear weak or absent.

This is a good reason to reconsider carefully such types of vision of our complexity.

The top–bottom mechanism is full of possible misinterpretations, and we know very well that the motor of all the systems is evolution. But evolution is active at the gene level, not at the level of ecosystems, or at the level of landscape matrices. No precise replies can be produced today without the introduction of a serious amount of biases.

The Mosaic as a Level of the Ecological Matrix

The mosaic can be considered as a level of the ecological matrix, the results of interactions between processes originated by organisms.

Mosaics are created by processes that interact at a specified level of complexity like the projection of a slide on a wall. To produce the image we need two things: the transparency of the projector screen and a surface on which to reflect the beam.

In most cases the mosaic that we observe is the overlap of many "projectors," and we can't distinguish the originating projectors, can't separate the different beams. This example could seem too simplistic, but, if you use this to continue the exploration of the nature of each mosaic, you can arrive at more interesting conclusions.

First of all, every projector could be a process and the wall a species. In this way, every species interacts with that process (beam) accordingly the different perception genetically driven.

Every species can be imagined at different distances from the projector intercepting more or less light and focused patterns. In this way we can introduce the scaling properties of each species and, instead of focusing a projector, we focus a species moving back or up.

We can translate this metaphorical representation of a reality:

1. Every species perceives the environmental complexity in a specific way.
2. Such complexity represents the context in which every species is embedded.
3. The level of definition of the flux of information that is originated from the matrix depends on the internal amplitude of the eco-receptors.
4. The eco-receptors cover all the functions of a species and can be associated with the vectors of the multidimensional niche.
5. The presence of a species close to another that has been recognized as a realized niche effect can be interpreted as an interference pattern.

Advantages of Living in a Mosaic-Like System

If a mosaic is a common pattern in the distribution of energy, resources, and organisms evolutionary and adaptive forces should be involved.

The advantages for a species of living in a mosaic may depend on:

1. The optimization of the resources that are distributed in patches.
2. The reduction of the hostility of unfavorable habitat (with patches of a mosaic being considered islands in a sea).

There are species that live in one type of patch and species that utilize a mosaic to live and to maintain different functions. We call the first specialists and the second generalists. But this is only one part of the story because there are species that use different patches to find the same type of food, and other species that collect a very heterogeneous category of food in the same patch. But considering different living traits and not only the feeding, for instance roosting, breeding, migrating and different periods of the year, we often can observe a very different reaction to environmental patchiness, and the number of cognitive mosaics intercepted by a species can be very high.

The border of functional patches is strictly regulated by the capacity of the biological sensor of species.

It is not easy to demonstrate that mosaics are patterns perceived by organisms and included in their genetic memory. Some examples are offered to discuss this interesting topic.

1. Many organisms cope with the mosaic-like pattern. For instance, animals living in flocks probably reduce the cost of individual anti-predatory behavior. Individuals that are at the border of the flocks may be the strongest, or may be males with redundant reproductive functions. The close vicinity of individuals such as in a shoal of fish or a pack of gnu encourages the exchange of information about direction and source of food or predatory vicinity.
2. The environmental mosaic is perceived by animals in different ways but many species have utilized the mosaic design to become mimetic.
3. Patches of suitable quality are often distributed at random. One good example is from the foraging behavior of white stork (*Ciconia ciconia*) by Johst et al. (2001) (Fig. 3.10). Studies in central Europe of this central-place forager have demonstrated the high sensitivity of white stork to the dynamics of foraging patches under human stewardship. Distance of patches from the nest, time after the mowing and asynchronous mowing are important components. The relationship between two different strategies of patch selection – by random or active decision in patch selection – is also important. Increasing the dynamics and the heterogeneity of mowing, we can expect an increase in patch-selection strategy against a random strategy. Strategies to manage patchy landscape can produce advantages for some species, for instance, changing harvesting practices from a simultaneous to a sequential mowing.

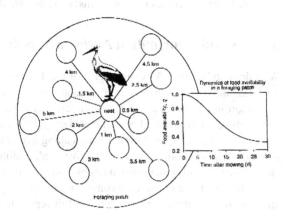

Fig. 3.10 Representation of spatial structure of landscape and availability of patches around the white stork nest. Food availability decreases following a logistic shape after mowing (from Johst et al. 2001, with permission)

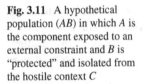

Fig. 3.11 A hypothetical population (*AB*) in which *A* is the component exposed to an external constraint and *B* is "protected" and isolated from the hostile context *C*

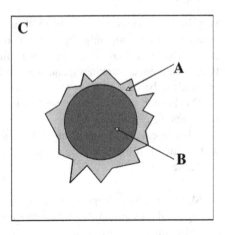

4. Patch position in a mosaic. At the border of the patches the uncertainty is higher than in the central part of the patches. Such uncertainty offers organisms new choices and the chance to become the driver of a system. But at the border the resources are more abundant and the competition is reduced.

Borders are for "pioneers," and the "frontier" is the place in which uncertainty is very high and far from self-organizing. In Fig. 3.11 we present a very simple model composed of three zones – A: interior, B: bordering, C: hostile. We can imagine that the flux of information in A is quite low if compared with B but system A is more stable than B and able to self-organize. In B the uncertainty is very high and the hostility of the matrix reduces the "attribute" efficiency that can be found in A. But in B individual variability can be used to cope with uncertainty and to find new strategies to expand and to dominate.

The thinness of a subpopulation shape is necessary to "penetrate" into the matrix by the occupation of small gaps existing in the hostile matrix.

The expansion front is composed of narrow propagules. This pattern can be observed either in plants (root systems) or in animal flocks.

Energy and Information Across a Mosaic

Mosaics can be created by the effects of local constraints that force crossing fluxes to divide into patches and to coalesce again and again.

According to this view a mosaic is created when a fluid crosses a medium in which there are some local constraints. This pattern can be observed in nature when we observe the gravel and sand deposits along a river after a flood. Water and debris are moved in a chaotic way along the river bed, finding local constraints like stones, abrupt curves, and human structures like bridges, basements, barriers, etc.

A clear example of such a mosaic is represented by the dynamics of a cloud. Air masses crossing the air matrix with well-differentiated constraints, such as wedges of cold air, act like attractors and create clouds that represent a type of mosaic in which every patch (cloud) after the constraint evolves toward a cloud formation.

Energy creates spontaneous mosaics of material differently spaced, and then such material can utilize additional energy internal to the system to auto-organize and to create different patterns.

In this way, a mosaic can be produced by the effect of energy in a highly dynamic material, like a fluid under locally spaced constraints.

In terms of information we can observe an increase in information moving from the center of each patch to the border where the information (uncertainty) reaches a maximum. This effect can be described in a very simple manner using a matrix in which two patches compose the entire mosaic. As you can see in Fig. 3.12, our mosaic is composed of two patches that have only one side in common, all the other sides are not considered. In other words, we have only the possibility to exchange information along the horizontal axis. In our model we consider the left and right side as the less informative for both the patches that meet the maximum of information at the center. We assume 0 as the value of the two columns placed respectively on the left and right end. We have not considered the upper and bottom sides. The reality is much more complicated, but this example may be useful to understand at least the behavior in one direction of the cells composing the mosaic.

The uncertainty based at the border of a patch can be incorporated in a system by adaptive mechanisms like the reinforcement of a cellular membrane and the increase of resistance to desiccation, to UV light, to animal browsing, etc. Genetic adaptation of organisms is very active at the border if compared with the interior.

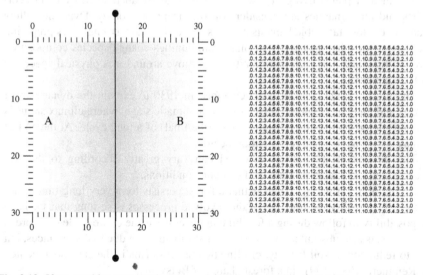

Fig. 3.12 If we consider every patch composed of n discrete entities (cells), it is possible to visualize that each cell has properties linked to position in Euclidean space. The closer a cell is to the border the greater the uncertainty. On the *left side* two patches are represented with an increase of darkness according to the increase of uncertainty. The *right side* shows the same image transformed into a numeric matrix

Developing the Mosaic Theory from a Plant Perspective

As often happens in the scientific realm theories are developed according to the approaches utilized. This is true in ecology in which the focal entities are plants, animals, virus, bacteria, fungi, or man. By using organisms most of our investigations are devoted to discovering the relationships between the specified entity and the environment.

We have much empirical evidence about the relationship between plants and the environmental mosaic as well as between animals and their habitats.

The separation of knowledge often creates barriers when we try to explore the complexity of ecological systems.

When projected into a map, plant distribution appears as a mosaic. The patchy distribution of most plants, especially if we are dealing with trees, is considered as a static pattern if observed at a short time scale.

Plants intercept the geo-morphological and climatic heterogeneity at multiple scales. Plants are extremely sensitive to environmental conditions and adopt several mechanisms to avoid hostile conditions or interspecific stressors.

Plants form complex communities that are distributed mostly through coping with the heterogeneity of the environment, and as a consequence, plant communities are distributed patchily everywhere.

For many decades, the dominance of the climax theory alone has restricted the role of spatial heterogeneity in the dynamics of plants. Today we can better understand the dynamics of plant assemblages from algae to redwoods.

The idea of plants living in a static system is today abandoned by plant ecologists and the dynamics are considered a rule in plant ecology. Dynamics allow plants to explore favorable habitats and to serve as involuntary engineers preparing the habitat for other species. That plant communities change species composition through time and that such communities can move around in a physical space are well documented.

The mosaic-cycle concept was developed in 1930 to explain the dynamism of forests and their successional stages toward a climatic stage, where climax status is not considered static but a transient phase (the final) of a long cycle. At the end of climax status, a complete rejuvenation occurs.

The dimensions of patches involved can vary greatly according to the soil heterogeneity and other climatic and edaphic conditions.

A common, yet wrong, belief is that a forest persists for a very long time if not disturbed. In reality the time lag of the involved processes is so long that we have no possibility to follow during a few human generations the entire cycle of a forest. Often forests are substituted by grassland or shrublands not due to disturbances, but due to reduction of soil fertility, or climatic changes. This is the case documented by Remmert (1989, 1991) in a forested area of Botswana.

Plants are extremely sensitive to geomorphological patterns that are quite fixed compared with other biological entities. For this reason, the pattern of plant distribution often appears static. With changing climate and/or the internal properties of the plant association, the mosaic expressed may change. This change has been called by various authors the "shifting mosaic steady state."

In reality, the changes that can occur inside a mosaic may be quite different and depend on the typology of vegetation considered. Grasslands and shrublands have quite different patterns than those of tree covers.

Grasslands and shrublands can modify the mosaic in a very short time and the agents involved may be soil arthropods, ground squirrels, moles, grazers, and diggers.

Trees (woodlands and forests) are influenced by at least three processes:

1. Development of vegetation along a successional gradient from the young stage to adult and senescent states. There is much evidence about this process common in every biome by which a succession of plants move in a direction of increase of respiration and decrease of productivity at ecosystem level (see Borman and Lickens 1979). The distribution of plant biomass changes along with the succession according to a cyclic fluctuation that can last a few months as well as many years (Fig. 3.13).
2. Gap dynamics. This process is linked to individual tree fall due to disease, senescence, attack by grazers, etc. The dynamic of gaps is very important in terms of regeneration in forests that do not have large disturbances, such as fires or hurricanes. The fall of an old tree has many effects on the soil and on the surroundings. For instance, the undergrowth close to the fallen tree suddenly receives light and is more exposed to desiccation. Several trees will incur damage to branches and competition for light and soil nutrients will be reduced. Around the gap an ecotonal area is created and this ecotone attracts new organisms like butterflies, birds, and new pioneering plants (Fig. 3.14).
3. Large-scale disturbance. The processes involved are fires, hurricanes, landslides, lava flows, earth-quakes, wind blows, climatic shock (severe frost, persisting drought). The changes that occur after these events depend on the typology of the forests, on the frequency of recurrence and on the season in which the event occurs.

In all cases plants can be substituted with other species, or a new generation of the same species might recover the open space (Fig. 3.15).

Fig. 3.13 Variation in biomass along an idealized patch. The increase of biomass is followed by a dramatic decrease for instance due to the death of a large tree. The decline occurs suddenly while the recovery takes longer (from Borman and Lickens 1979)

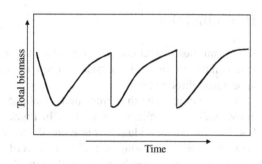

Fig. 3.14 The gap dynamic
in a forest may produce a
rejuvenation of the same
species (A), or new species
can enter into the forest
allowing the empty gap to
recover (B)

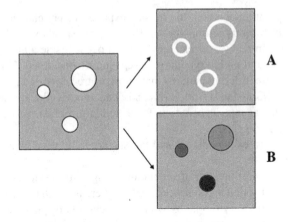

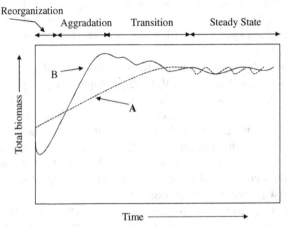

Fig. 3.15 Two models of patch development after a disturbance. *Line A* is an asymptotic model in which biomass accumulates until the Steady state. This model is well known and accepted by plant ecologists. *Line B* describes the Shifting-Mosaic Steady State proposed by Borman and Lickens (1979) in which after a phase of reorganization the aggradation phase grows quickly followed by a transition phase in which biomass decreases to enter into the final phase of the Steady State (from Borman and Lickens 1979)

Discontinuities

Discontinuities in soil composition, plant cover, resources, and individuals are a common pattern in every ecosystem. These discontinuities change in extension and appearance through time.

So we can imagine that every individual perceives every unity of the perceived environment as hospitable or hostile. In Fig. 3.16 a model of hospitality versus hostility for an individual is represented. For instance, in many plants the perception of the neighboring area changes according to season, age, and community composition, consequently changing the model.

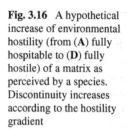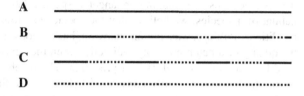

Fig. 3.16 A hypothetical increase of environmental hostility (from (**A**) fully hospitable to (**D**) fully hostile) of a matrix as perceived by a species. Discontinuity increases according to the hostility gradient

Discontinuities can be observed also at a scale that is larger than the individual-based scale. In this case discontinuities are transformed in patchiness. Moving from discontinuities perceived by individuals at a larger scale, such discontinuities are perceived as a mosaic of favorable or less favorable conditions. In this case the populations are interested in such a process.

In 1988, Pulliam presented a model on distribution of small birds across an environmental continuum (Pulliam 1988, 1996). He recognized that some populations that he called "source" had a positive balance between births and deaths and emigration exceeded immigration (Fig. 3.17). Under other conditions some populations had an opposite trend in which the number of births was not sufficient to balance the number of individual deaths and immigration exceeded emigration. He called this type of population a "sink."

In a population of source type, the surplus of individuals moves to neighboring spaces, including also less favorable habitats. The populations of sink type could go extinct in a short time if an active immigration of individuals from a source population does not balance the individuals that have perished. This continuous flux of individuals from a more favorable to a less favorable habitat is fundamental for the persistence of populations. The Pulliam model is really important to understanding the behavior of populations under different environmental constraints and opens up

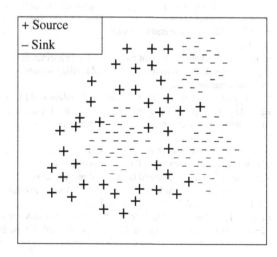

Fig. 3.17 Source-sink model of a population in a patchy habitat

new perspectives in managing endangered species. When we are dealing with the habitat of a species, we believe that the species has found the best environmental conditions. Pulliam, working on sparrows, has demonstrated that species recognize the quality of a patch probably indirectly, when the survival of individuals is menaced by starvation or high risk of predation. The patchy nature of the habitats (in sensu stricto) was not fully considered before the Pulliam model. A gradient of quality for habitats is the rule and not the exception in nature and habitat-specific demography plays a relevant role in regulating natural populations.

According to the Pulliam demographic model (see Table 3.1), Lambda is the finite rate of increase for populations and appears positive in the source habitat and negative in the sink habitat.

The source-sink dynamic is de facto the confirmation that patchiness operates not only on vegetation patterns but also at the level of animal populations. Along a suitable period of time, it is possible to observe in the same habitat populations of source type and in other years populations with sink characteristics. In order to evaluate in a determined time lag if a habitat has a positive or negative demographic trend, Pulliam suggests the calculation of the geometric mean of all $(\lambda 1 \lambda 2 \lambda 3 \ldots \lambda t)1/t$. If the mean of λ on a long period is <1, the population declines; on the contrary, if $\lambda > 1$, the population has a net gain.

The model proposed has been further implemented by Watkinson and Sutherland (1995), who introduced the concept of pseudo-sink. According to this model, some populations considered sink can maintain a minimum population also in the absence of migration, and this is a further confirmation of the complex dynamics of populations in patchy habitats.

In many cases some patches are true ecological traps for species that are attracted by some factors like food, roosting availability, nesting suitability, but other factors like disturbance, predation, or competition can produce a negative balance in the

Table 3.1 Demographic source-sink model (from Pulliam 1988, 1996)

---

n_T = Individuals at the end of winter

B = Offspring

$n_T + \beta n_T$ = Individuals alive at the end of the breeding season

P_A = Adults that survive during the nonbreeding season

P_J = Juvenile survival

$n_{T+1} = P_A n_T + P_j \beta n_T = n_T(P_A + \beta P_J) = \lambda n_T$ Population at the beginning of the next year

Lambda = $(P_A + \beta P_J)$ is the finite rate of increase for the population

In a heterogeneous environment habitats may sustain source or sink populations with different levels of $\lambda 1,2,3$

If $\lambda 1 > \lambda 2$ then $\lambda 1$ is source and $\lambda 2$ is sink

$n1*$ = is the maximum size of a source population

$\lambda 1 n1*$ = Individual at the end of the nonbreeding season

$\lambda 1 - 1$ = per capita reproductive surplus in the source habitat

$\lambda 2 - 1$ = per capita deficit in the sink habitat

$n1*(\lambda 1 - 1)$ = Number of individuals that migrate in sink habitats

$n2* = n1*(\lambda 1 - 1)/(1 - \lambda 2)$ Equilibrium in a population of type sink

---

population. In this case, we can observe a denser population in such habitats that in reality is the result of immigration fluxes from other source habitats.

The source-sink model opens up new perspectives on the mosaic theory indicating that often the local density of a population is a misleading indicator of habitat quality. The seasonal variation in habitat quality for a species can determine characters of source in one season and of sink on another occasion. This source-sink patchiness allows better understanding of the dynamic of populations, especially over a long period. Some good habitats can be found empty while other habitats that are considered poor habitats can be found to be full of organisms. In conclusion the demography of populations is quite far from being understood in detail, and the role of patchiness could be more important than suspected in the past.

Complexity of the Ecological Mosaics

Often we are thinking about complexity as a status of the matter in which the relationships are the main constraints or actors. This is correct, but it is not the only condition to explain the observed complexity.

Complexity means capacity for self-organization, irreversibility, cybernetics, but it means also opportunity for the appearance of new processes, patterns, and biological forms.

Stochasticity, or if you prefer information, must flow easily, at the edge of chaos, to create new conditions.

For this it is reasonable to hypothesize – in addition to connectivity between the components of a system – independent mechanisms able to isolate temporarily some parts of a system, to organize such parts and then to fragment again such parts to form "creative" opportunities and to insert such parts into the processes.

Complexity in the mosaic represents the rule and not the exception. Such complexity depends mainly on the nature of the mosaics and on the relationship with the organisms that intercept such figures.

Complexity in a mosaic means that a mosaic is composed of nominal juxtaposed patches like a puzzle, but in turn every patch can be modified at any time by internal or external mechanisms, creating a dynamism that at the end influences the structure of the whole mosaic.

The Maintenance of Patchiness Inside a Mosaic

Patchiness is a patterned character of a mosaic and is represented by the juxtaposition of patches of different types. A patch is a discrete element of the mosaic (the unit of a mosaic), and it is created and maintained by internal as well as external constraints. The internal constraints that we call centripetal processes act to incorporate energy, matter, and organisms from the outside against the gradient. In this way a patch is like a biological cell with a homeostatic capacity to isolate from the external environment. The internal processes need energy to use neg-entropy and produce a catabolism that can be beneficial for neighboring patches.

Coupled with this centripetal process, a centrifugal process moves energy, matter, and organisms in the opposite direction as the product of the patch dynamic. In this way, seeds, young animals, and nutrients flow out from the source, dispersing in the outside environment.

We can imagine a patch in a status of pulsing during which at each contraction or release an active circulation of energy, matter, and organisms is carried out in both directions. A patch can be considered an importer, but also an exporter, and this activity of import/export assures dynamism to the entire system of which a patch is part. At the border of each patch, ecotones create a border like a cellular membrane.

Self-Organizing of Mosaics and the Fifth Dimension

In a simplified view a mosaic can be considered "simply" as the combined action of environmental constraints that in turn create a complex system in which patterned sub-systems interact. But in reality, a mosaic is much more than a deterministic and "Euclidean" or "fractal" process of patch juxtaposition, or simply a three-dimensional space in which we recognize objects. In fact although we add another dimension (for volumes), resulting in a fourth dimension, this is not enough to jus-tify the complexity of the system under evaluation (Fig. 3.18). We need to add another nonEuclidean dimension (a process dimension) that we call simply the "fifth" dimension (Fig. 3.19). This dimension probably contains the "overlap" of mosaics produced by different and/or independent processes. The fifth dimension is fundamental to the structure of the ecological complexity, and as in the niche theory, causes the emergent character of the biological complexity to appear. This dimension can be considered a meta-domain in which several "mosaic" domains

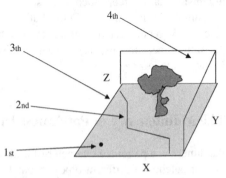

Fig. 3.18 The first four dimensions of a mosaic are represented respectively in Euclidean space (x, y, z) by point, line, space, and volume

X,Y (first dimension) (point)
X&dX (second dimension) (line)
X&Y (third dimension) (space)
X&Y&Z (fourth dimension) (volume)

Fig. 3.19 The fifth dimension (or mosaic meta-domain) is created by the contemporary presence of different mosaic-patterned processes inserting into the same topological space

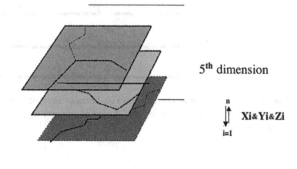

5th dimension

$\sum_{i=1}^{n} Xi \& Yi \& Zi$

interact producing emerging properties. The fifth dimension is a multiprocess space in which every process can act without affecting primarily all the other processes.

In the fifth dimension different mosaics are in action contemporarily exchanging information primarily inside the focal mosaic level but other emergent properties arise. We have to eliminate the misconception that all parts are linked to all other parts, substituting this assumption with another: some parts are more connected then others, and some connections are ephemeral. This assumption is quite important to understand how a complex system functions. We are used to considering an ecosystem as a full connection of elements that exchange information, energy, and material. This is true in principle, but, if we try to measure the strength of such connections, it is easy to verify that some connections are two or more orders stronger than others that can be considered marginal (Fig. 3.20). In conclusion, the overlapping of different mosaics into the fifth dimension is allowed by the fact that each mosaic moves scarce information from and to the other mosaics. One fundamental factor that allows such a dimension is represented by the time lag of the processes, the recurrence of phenomena and the interval by which a process is modulated (Fig. 3.21).

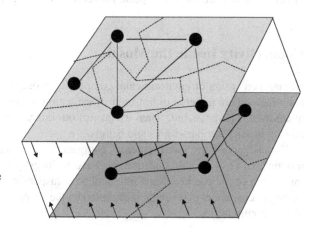

Fig. 3.20 The connections inside a mosaic are two or more orders stronger than the connection between the different mosaics of the fifth dimension

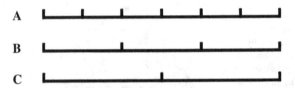

Fig. 3.21 Different mosaics can coexist because the focal processes have a distinct temporal (scale) frequency. In this case mosaic (**A**) has six recurrences, mosaic (**B**) three, and finally mosaic (**C**) two recurrences

A little more explanation of ecological time is necessary because there is a risk of confusion on terminology.

Time is a physical parameter, and we can't change this, at least in this universe, but in the same time lag the process can behave differently. The speed and frequency of a process is the way in which a process repeats a sequence in an interval of time.

Strong and weak connections between parts are widespread in the fifth dimension. The weak connections operate at the margin of the principal nodes that link the different patches.

There are in conclusion three ingredients of mosaic complexity: variety of internal and external constraint, weak connections between the overlapping mosaics, and different patterning in space and time. For instance if we use such a paradigm to shape a real mosaic in the field, we can produce many mosaics from the same image. What can be the meaning of this phenomenon? Probably some mosaics extracted by the aerial image (see f.i. Fig. 3.4) are not working in reality, but most of these can be coupled with a process or an organism. For instance, the mosaic represented only by two patches indicates the soil water content, the last with many patches the plant community.

The fifth dimension also incorporates the hierarchical organization of a mosaic. The hierarchy as later illustrated is another type of organization in which space is more relevant than time and in which elements are composing parts of a major system, and this sequence is repeated several times along the hierarchical chain.

Connectivity Inside the Mosaic

Human perception of environmental complexity is only a partial part of the reality and often we are tempted to limit our investigation to our scale. This could be adequate for our purpose, but, when we project our interest beyond our perception, we have to modify our paradigms and beliefs.

Which part of the environmental complexity is created and maintained by a mosaic? To reply to this question we have to investigate the role played by the interaction of the patches. Such interaction (intraspecific or interspecific) produces emergent characters that can be intercepted by the species, or allow the organization of new communities.

The intraspecific mosaic is represented by patches of the same type that are separated by patches of other types. Every patch is not in direct contact with all the others but is influenced to a certain extent by their distance. In landscape ecology we call this effect connectedness. In reality connectedness is a measure of a pattern (the physical distance between patches), and it is not a process. On the other hand, connectivity is the measure of distance between patches when these are calibrated by processes or organisms.

Connectivity is an emerging property of every mosaic when intercepted by an organism. We will discuss this in greater detail where the paradigm of the ecotone will be described.

Connectivity may be an important element for a species living in a mosaic and meta-population theory approaches this problematic. Generally, meta-population theory refers to mobile animals like butterflies and birds.

When we are dealing with plants, the distance between patches composed of the same species can play a fundamental role in keeping the local population in good shape by cross-breeding and thus avoiding inbreeding.

The mosaic structure assures a high level of dynamic for organisms and their aggregations and allows energy to cross different levels of the hierarchical structure degrading in entropic processes or upgrading by neg-entropic processes.

Suggested Reading

Farina, A. 2000. Landscape ecology in action. Kluwer Academic Publishers, Dordrecht.
Zonneveld, I.S. and Forman, R.T.T. (eds.) 1990. Changing landscapes: An ecological perspective. Springer-Verlag, New York.

References

Borman, F.H. and Lickens, G.E. 1979. Pattern and process in a forest ecosystem. Springer-Verlag, New York.
Golley, F. 1993. A history of the ecosystem concept. Yale University Press, New Haven.
Hutchings, M.J., John, E.A., and Stewart, A.J.A. 2000. The ecological consequences of environmental heterogeneity. Blackwell Science, London.
Johst, K.J., Brandl, R., and Pfeifer, R. 2001. Foraging in a patchy and dynamic landscape: Human land use and the white stork. Ecological Applications 11: 60–69.
Kolasa, J. and Pickett, S.T.A. (eds.) 1991. Ecological heterogeneity. Springer-Verlag, New York.
Kolasa, J. and Rollo, C.D. 1991. 1. Introduction: The Heterogeneity of heterogeneity: A glossary. In: Kolasa, J. and Pickett, S.T.A. (eds.), Ecological heterogeneity. Springer-Verlag, New York, pp. 1–23.
Levin, S.A. 1976. Population dynamic models in heterogeneous environments. Annual Review of Ecology and Systematics 7: 287–310.
Pulliam, R. 1988. Sources-sinks, and population regulation. American Naturalist 132: 652–661.
Pulliam, R. 1996. Sources and sinks: Empirical evidence and population consequences. In: Rhodes, O.E., Chesser, R.K., and Smith, M.H. (eds.), Population dynamics in ecological space and time. The University of Chicago Press, Chicago, pp. 45–69.
Remmert, H. 1989. Okologie. Springer-Verlag, Berlin.

Remmert, H. 1991. The mosaic-cycle concept of ecosystems – An overview. In: Remmert, H. (ed.),
    The mosaic-cycle concept of ecosystems. Springer-Verlag, Berlin, pp. 1–21.
Tansley, A.G, 1935. The use and abuse of vegetational concepts and terms. Ecology 16: 284–307.
Watkinson, A.R. and Sutherland, W.J. 1995. Sources, sinks and pseudo-sinks. Journal of Animal
    Ecology 64: 126–30.
Wiens, J. 2000. Ecological heterogeneity: an ontogeny of concepts and approaches. In: Hutchings
    M.J., John E.A., and Stewart A.J.A. (eds.), The ecological consequences of environmental
    heterogeneity, Blackwell Science, Oxford, pp. 9–31.

Chapter 4
Properties of Ecological Mosaics

Introduction

The mosaic appears everywhere and is considered to be an aggregation of basic heterogeneity, but, if we reduce or enlarge our vision, new patterns appear. Moving down, we expect to observe the appearance of heterogeneity. But what is the limit at which we establish that a system is simply heterogeneous? I consider this a poor question; heterogeneity appears and disappears according to the scale of observation. We can state that heterogeneity appears many times as we move across a changing scale, and the same happens for the mosaic.

In Fig. 4.1a hypothetical model is presented, in which mosaic and heterogeneity appear alternating with movement across different scales. This alternation can be perceived only under the condition that we change the scale of observation. We have to change either extension or resolution, and this is possible only if we imagine use of the sensors of different organisms. Often we maintain one of the two parameters or we consider only the human perception of our environment.

Mosaic and heterogeneity have fractal behavior, and this allows measurement of the complexity of "mosaics" and "heterogeneities" that are present across scales.

The investigation of the properties of a focal mosaic could be developed using fractal analysis and applying the indices of complexity to the entities that appear when we change the extension and resolution of the images. Considering j a value of extension and i a value of resolution we can measure the level of heterogeneity and patchiness at ji scale.

Heterogeneity reaches a maximum when all the categories have the maximum spatial invariance and a minimum when the spatial variance is at a maximum. Patchiness is at a maximum level when the contagion (see later in the chapter 7 on methods) is at a maximum (Fig. 4.2).

Hierarchical Organization of Ecological Mosaics

It is a priority to link and to integrate information from the structural analysis and the species-specific perception of mosaics.

A. Farina, *Ecology, Cognition and Landscape*, Landscape Series 11,
DOI 10.1007/978-90-481-3138-9_4, © Springer Science+Business Media B.V. 2010

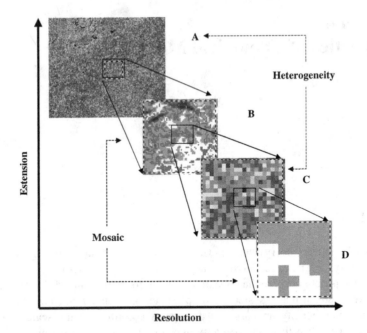

Fig. 4.1 Mosaics appear at every level of resolution alternated with heterogeneity. In cases *A* and *C* the heterogeneity is the dominant pattern and it is not possible to observe mosaics. In cases *B* and *D* the mosaics are dominant. *A* and *C* are considered disordered images in which it is not possible to observe patterns linkable to explicit ecological processes. In *B* and *D* the ordered vision dominates and if we consider such mosaics created by plant cover we can address the question: what were the factors responsible for such patterns?

When we analyze ecological mosaics, it is clear that, according to the scale utilized to observe such forms, different figures appear. Often this is possible, especially for plants, but this analysis can be carried out also using animals to distinguish sub-systems inside each mosaic.

It is reasonable to imagine that when we can observe two different types of patches inside a mosaic, it is possible to extract more detail on each patch type and distinguish another more fine-grained level inside each patch. This character of patches can be explained in terms of the hierarchical theory.

Well argued by O'Neill et al. (1986) hierarchical theory is useful to interpret the complexity of ecological systems, especially when we try to investigate the

Fig. 4.2 Hypothetical table of contingency between extension and resolution measuring the value of heterogeneity and contagion considering paired changes in resolution and extension. An alternate highest score for heterogeneity and patchiness is expected

	Estension			
	J1	**J2**	**J3**	**J4**
I1	H>C			
I2		C>H		
I3			H>C	
I4				C>H

Resolution

H=Heterogeneity   C=Contagion

mechanisms that create self-maintaining structures. At every level of hierarchy we observe sub-systems that are created and regulated by specific processes.

I hope the example I report is clear enough.

At Cerreto Pass, in the Northern Apennines there is a chalk evaporitic deposit from the Mesozoic era. The surface of this deposit has been modified by karstic processes creating dolines. Focusing on the forms of the relief, we observe different types of dolines, distinguishable by size, shape, and depth (Fig. 4.3).

Such dolines are the result of dissolution of chalk, and the mosaic created by such a process produces steep micro-relief. If we move in to a finer scale, investigating the dynamics inside each doline, the scenario appears more complicated, and to the geomorphological processes we have also to add biological processes that act on soil, modifying the dissolution rate of the chalk and the entire hydrological regime.

Moving further inside such a mosaic and focusing on the vegetation a completely different scenario appears. In fact it is possible to distinguish between tall shrubs and open prairies. This mosaic is the result of human intervention over a period of more than 5,000 years. Human activity has created clearings used for livestock grazing or for cultivation since protohistorical time.

In the past such clearings were probably abandoned during periods of famine, disease, or war, but again and again have been maintained by different degrees of disturbance. Such types of stewardship have so deeply influenced the landscape that today all processes linked to biological components are affected by this footprint (or memory).

So in conclusion, looking at vegetation on a broad scale, we can distinguish two types of patches: open areas and shrublands. Both types are the result of a long-term regime of disturbance through human use.

The mosaic created by geomorphological processes (doline) is independent of this last process created by humans, but moving up the scale something changes rapidly.

Moving inside the open prairies created by human intervention and changing the scale of resolution, we can observe another mosaic composed of different types of herbaceous plants. It is possible to distinguish between Ericaceae, *Brachypodium pinnatum*, and other grasses. Such a mosaic is largely affected by the grazing pressure of domestic (cows, horses, sheep, and goats) and wild grazers (Roe deer, hare, and insects).

A fourth level can be discovered inside each prairie patch type, and again we can distinguish different levels of standing biomass. The process involved is represented by soil fertility, in turn created by input of organic material (urines, feces) or by the activity of soil microbes and fungi.

The relationships between the different levels are not direct, and this is the main source of complexity. The mosaic is created by independent processes acting in the same place at different time lags.

According to the hierarchical theory, every system is the result of the contribution of concurring sub-systems, and this mechanism persists across different levels of spatial and temporal scale (Fig. 4.4).

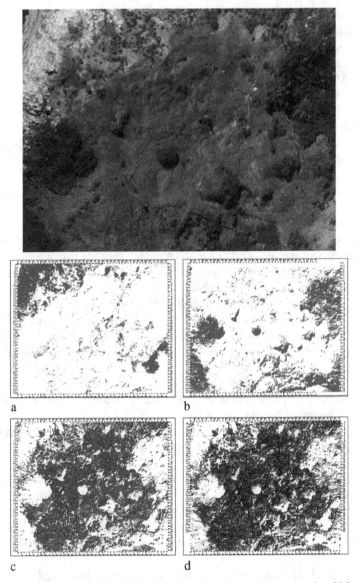

Fig. 4.3 Example of a scaled image interpretation of a sub-montane pasture in which geomorphology (**a**: karstic processes), land use (**b**: open areas, shrublands), and fine-grained use of the vegetation (**c**: plant composition, **d**: standing biomass) create a complex mosaic. See text for explanation

The dynamics of a system depend on the hierarchical rank of the system: High-rank systems have slow dynamics. For instance, a biome changes over the long term while the plant community of our garden can change completely in one season. The different speed at which different hierarchical mosaics modify their structure is a fundamental force contributing to the overall complexity (Fig. 4.5).

Fig. 4.4 The transition from one level to a level of higher rank produces emerging characters from a filter zone

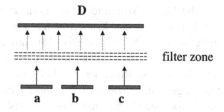

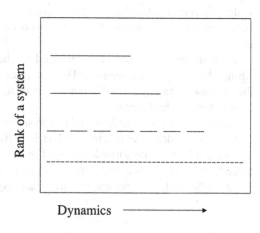

Fig. 4.5 Dynamics increase as one moves down the hierarchical structure of every system

Scaling Properties of Mosaics

The dimension of patches (the mosaic grain) depends on the object composing the patches and on the disturbance regime.

Generally, the increase in disturbance increases patchiness and reduces the scale at which living forms organize in the environmental context.

In this way the anthropogenic landscapes of the recent past have been found by Perevolotsky and Seligman (1998) to be more fine grained than the present-day landscape after land abandonment and shrub and woodland recruitment. Patch dimension is strongly affected by the disturbance regime and by its consequences. In the Mediterranean mosaic the finer scale is richer in biological diversity than a coarse-grained scale created by shrub and tree recovery. In conclusion, the grain of a mosaic may be a good estimator of biological diversity.

Agents of Temporal Changes of the Mosaic Structures

In the dryland of Zimbabwe, the land mosaic is strictly linked to many environmental as well as human factors. Biomass distribution is highly variable in space, and unpredictable according to the season (Scoone 1995).

The habitat structure at any point depends not only on the surrounding habitat but also on its history, magnitude, and the trajectory of changes at multiple scales. The hypothesis that animal distribution in a shrubsteppe habitat depends not only on current landscape but also on habitat change was tested by Knick and Rotenberry (2000) in 200,000 hectares of Southern Idaho. In this region the invasion of *Bromus tectorum* by overgrazing and failed agricultural homesteads has increased the risk of fire affecting the community of big sagebrush (*Artemisia tridentata*), winterfat (*Kraschenninikova lanata*), and shadscale (*Atriplex confertifolia*). The frequency of fire has increased over the last years, fragmenting and reducing the shrub cover from 51 to 30% of the total area.

Birds in particular seem to be affected by these changes although they show a strong philopatry or site tenacity. The use of satellite images allows us to establish the increase of reflectance of about 33.8% of the scanned pixels as a consequence of wild fires in shrubsteppe.

Disturbance plays a fundamental role in shaping landscape distribution of vegetation as recently demonstrated in Mt. St. Helens 20 years after the eruption (Lawrence and Ripple 2000). The variables utilized in this study were: disturbance type, distance from crater, tephra thickness, blast exposure, distance from surviving forests, slope gradient, slope curvature, elevation, and aspect.

Grazing by Large Herbivores

Grazing is an important process in shaping a landscape. Grazers have a quite complicated multiscalar approach according to the different positions of landscape, selecting plant communities, feeding stations (patches), plants, and plant parts. Grazing modifies the structure of grasslands. Sparse sward yields larger bites and leads to a more rapid depletion than short dense cover (Laca et al. 1994).

For instance, bison interact with the patch structure of grasslands at several spatial and temporal scales. Grazing reduces the standing biomass while vegetation diversity is increased as a consequence of the intermediate disturbance hypothesis.

Patchiness is often the result of different combinations of disturbances; for instance, fire+grazing+urine deposition. It is well documented that bison, like many other large grazers, have a preference for patches created by urine deposition. The urine deposition is a disturbance of small size (≈ 0.25 m^2), but Steinauer and Collins (2001), using experimental control plots sprayed with bison urine, have found an increase in size and severity of grazing for patches with urine deposition. In conclusion, grazing is initiated in a urine patch that is very attractive to bison due to an increase in biomass productivity and quality (through N addition to soil nutrients) and this has consequences for the total grassland mosaic.

Windthrow Disturbance and Mosaic

Many studies have emphasized the role of abiotic factors in forest dynamics. Fires, catastrophic windthrow, drought etc. are all important elements in steady-state and gap-phase-dominated models. Little information is available on the consequences of these stressors in terms of rate, scale, and severity across the landscape.

Among the disturbances that modify old growth forests of the Pacific region from Canada to Alaska, the role of the gap-phase model has been overestimated. Recently Kramer et al. (2001) collected evidence about the role of windthrow in shaping temperate rain forest covers in Southeast Alaska. This disturbance is particularly evident on windthrow-prone slopes in which large-scale perturbation is dominant. In storm-protected areas, gap-phase processes are dominant, and it is in these areas that most of the timber harvest is concentrated. Slope, elevation, soil stability, and exposure to prevailing storm winds seem to determine the patch dynamics of these forests. The windthrow return interval has been estimated at 300 years, and late seral stages are maintained especially in wind-protected areas in which the gap-phase dynamic is dominant. Knowledge of the dynamics of these forests is of great importance for long-term management. The removal of standing biomass in wind-protected areas modifies the scale of disturbance from the micro to large scale with unpredictable effects.

Fires and Mosaics

Fire is one of the most important agents of change across a landscape. Fire frequency and severity can modify the structure and the spatial distribution of patches. Land managers and ecologists have become involved in a significant debate related to the management of fire in order to mimic natural disturbance regimes. According to different biomes, fire recurrence can be from a few decades to 400 years or more. Human intrusion often increases the frequency of recurrence producing dramatic changes in vegetation patterns, although this is not the rule as demonstrated recently by Floyd et al. (2000) in the Mesa Verde National Park in Colorado (Fig. 4.6).

Patch Dynamics and Animal Responses

The dynamics of a mosaic may depend on many factors (see for instance Levin 1976). Recently, the theme of patch dynamics, especially fragmentation, has been attracting the interest of ecologists. Patch dynamics can be experimentally duplicated and tested on natural populations. For this not only fragmentation per se but also network design and patch corridors are under study through the use of different groups of plants and animals (Fig. 4.7). For instance, Summerville and Crist (2001) in an old field tested the effect of induced fragmentation on the butterfly population (see Fig. 4.8), finding a direct effect of treatment on the populations.

Fig. 4.6 Map of the history
of fires for the petran
chaparral vegetation in Mesa
Verde National park in 1998
(from Floyd et al. 2000)

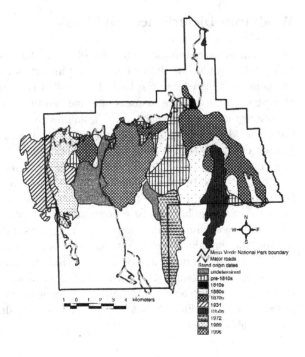

Fig. 4.7 The interception of
moving organisms in a
continuous landscape is
higher than in the fragmented
one

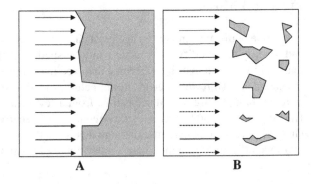

Community and species were linearly related to the remaining habitat according
to the proposed model. Plots with a higher number of forbs were visited more than
poor plots demonstrating independence between patch structure and quality of patch
per se. This could suggest that even modest areas can provide enough resources of
quality if managed in terms of richness and may be equivalent to larger patches
of modest quality. Species respond differently to fragmentation. Rare species are
especially sensitive to fragmentation but other species seem not to be affected by
fragmentation treatment.

Fig. 4.8 Histogram showing the effect of fragmentation treatment on a butterfly community. Rare species are present in large numbers only in unfragmented patches (15 × 15 m) of the experimental area (from Summerville and Crist 2001)

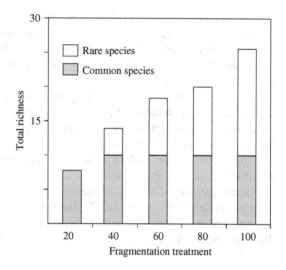

Resilience and Other Properties of Ecological Mosaics

Mosaics are composed of patches of different type, size, and shape. Theses differences on one hand create problems of isolation, fragmentation, and edges for many species, but on the other hand reduce connectivity for the spread of diseases, species invasion, and disturbances.

A patch per se can benefit from fragmentation and isolation depending on whether it is a population patch, a community patch, or a process patch.

Patchiness increases the difficulty for prey patches to be intercepted by predators, and reduces the risk to interception by a disturbance like fire, a wind storm, or simply by a local tree fall. On the other hand, fragmentation increases the number of ecological traps and can be very attractive for many species that assume the character of population sinks.

This resilience is a property of many systems and can be defined as the capacity to incorporate disturbance. Every system has resilient properties, but such properties can be reduced by stresses like pollution or climatic changes.

Resilience means the capacity to recover after disturbance. This attribute of the majority of environmental systems operates differently in a mosaic than in an ecosystem. A mosaic can be described as a spatial system in which a set of organisms self-organize complex systems around every discontinuity. The self-organization requires a low level of heterogeneity in order not to consider such heterogeneity as a stressor per se on the mechanisms responsible for the self-organization. A mosaic can be considered a real system in which patches represent parts of the entirety. For instance, the upland prairies of the northern Apennines, above the tree level, for their relatively simple vegetational structure are wonderful laboratories to experiment with mosaics.

A hierarchy of stressors of shaping factors is responsible for such a mosaic. The highest rank is dominated by the montane climate (oro-Mediterranean). Second are the aspect and steepness that in turn control light, temperature, wind effect, snow cover, rain distribution, and fire risk. In third position are the soil properties which are linked to the geology.

The fourth system is determined by water content in the soil which increases moving from top to bottom. The fifth component is represented by livestock disturbance which is concentrated on the southern aspects and also all around the top of the mountain. Every factor acts at a different spatio-temporal scale. At local scale (micro-site) the neighbor effect around the center plant occurs to determine the fate of individual plants.

Habitat patchiness assures resilience to human-modified systems, as pointed out by Scoone (1995) in dryland Zimbabwe for livestock living in heterogeneous habitats. The phenology of resources allows livestock to find different resources in different periods of the year.

Temporal Characters of Ecological Mosaics (Ephemeral, Seasonal, Multiyear)

Time is an important constraint for every ecological mosaic.

Short-term mosaics, also considered ephemeral, can be created by a seasonal rainfall, by the melting of snow masses in late spring, by the frozen lakes of winter, or by a flowering of algae in a pond. Such mosaics are considered ephemeral, but despite their name often contain rare or endangered species, or are essential for maintaining populations of common species like amphibians.

Many ephemeral mosaics contain a rich flora or fauna or a mix of the two. The time scale is the only distinction possible across mosaics. Species that can interact with ephemeral mosaics have a life cycle synchronized with the time scale of the processes responsible for that mosaic.

The advantage for a species of interacting with such an unpredictable and short-term mosaic can be the use of suddenly appearing resources, or due to the unpredictability, a way to escape predators or to avoid competition. Predation by fish is not possible in temporary ponds, and in this way amphibians escape such types of predation. Also terrestrial predators like snakes need time to locate temporary ponds.

Most terrestrial amphibians use temporary ponds for egg-laying and tadpole development. The reproductive cycle of evolutionarily ancient organisms is completed in a short time in an ephemeral patch!

Mosaics produced by fires are a special case of plant growing. Many dormant seeds can be stored in the soil for a long time, and their development is managed by fires that create clearings in the forest cover.

In this case the mosaic is the product of a process that at the end of the story allows the arrangement of plants in the same patch.

Patterned and Process Mosaics

The distinction between patterned mosaics and process mosaics depends largely on our capacity to intercept each category. The patterned mosaic, like the ones created by plant association, can be detected by our senses, although we perceive only one part of the complexity created by such mosaics.

The lunar landscape is without life; no organisms could enter into contact with such a mosaic. Only human technology allowed the landing of astronauts, but again the astronauts were confined to an earth atmosphere, the only contact with the moon landscape was visual. The same story could be experienced if, applying fantasy, we could imagine a trip to the center of the earth moving within fluid rocks.

The second category, the process mosaic is intercepted only by species-specific perceptional and cognitive mechanisms.

The economic balance in a rural society can not be easily observed using the vegetation mosaic, but nevertheless richness or poverty, abandonment or intensification in land use are the effects of economical processes that can not be detected directly (for instance using remote sensing procedures).

At this point it appears important to spend few words on these two types of mosaic:

1. The patterned mosaic, always detected by tools like the remote sensing technique (vegetation indices, aridity, biomass, plant diseases, etc.). This is the mosaic that exists over a broad range of scales and physical and biological domains. This is the mosaic that we perceive directly using our senses. It has been such perception that for a long time has created segregated disciplines each devoted to the study of a component of such mosaic; for example geology, stratigraphy, vegetation sciences, etc. Human culture often fears approaching this in a more holistic way. The holistic vision rejected by reductionistic science represents a short cut to perceive the complexity as an emergent phenomenon common to physical and biological worlds.
2. The process mosaic. This mosaic can be detected only after the interpretation of the effects produced. Such a mosaic, for example economy and environmental health, in effect represents the highest level of organization of complex systems. In this category we can include the cognitive mosaic created by the knowledge of an individual of the used surroundings.

So dealing with economic or cultural landscapes is not a simplification of the reality but a reality "tout court." In fact economy and culture represent emergent processes that transcends the individual habits but in turn produce changes on the single components of our society. Both require a great investment in research because their detection is not immediate.

Culture, for instance, is moving at a different velocity than economy and often represents a social buffer or, in some cases, a container or resistance to novelty.

At this time, the economy seems to be the emergent driver of social, political, and cultural compartments.

The mosaic paradigm helps reductionists open their mind to complexity as an emergent character of the real world.

There are two main ways for an organism to perceive the complexity: the first mechanism is direct perception of the heterogeneity in which a species is living. Such heterogeneity largely depends on the eco-field of focal organismic functions (see Chapters 8 and 9).

The second way is represented by the effects or constraints created by some processes that reduce the degree of freedom that such organisms have. De facto this reduces the "potential" fitness of the individual.

One such constraint is represented by intra and interspecific competition. Competition appears only when some conditions occur in the context in which a certain individual is living. Competition may be linked to available resources in the soil, light, or water for plants, or to food or territory if animals are considered.

Patterned mosaics have strong and direct scaled consequences on species, and represent the footprint in which most of the organismic functions are related. Such footprints play a fundamental role in species distribution, abundance and, more generally, individual fitness.

The process mosaics can be located in a second row in terms of importance and relationship. Competition becomes important, but the pre-requisite is represented in the case of a plant by the presence of biological soil.

Scaling the Mosaics by Species

It seems that three contexts are possible:

The first context is one in which the presence of an organism is conditioned by favorable conditions like water availability, soil fertility, and micro-climate. This seems quite a "passive" status in which an organism is enveloped. For a plant this could be the seed germination after dispersal from the mother plant; for an animal, the selection of a territory in which to reproduce.

A second step is represented by the direct interception of the species-specific eco-field in the context. This step depends greatly on the species considered. For a plant this could be the development of a root system or the shape of the canopy. Both are largely affected by availability of nutrients, by light tenure, by the fungi community in the soil, or by aerial competitors.

For an animal the second step can be represented by the availability of enough food, good reproductive sites, or low interindividual competition.

The third context, which represents the process mosaic, is again like the first one strongly dependent on the context and passively "accepted" by organisms. The economic mosaic can create patches in which timber logging assumes a particular importance for local populations and this in turn favors the reduction of old growth forests. It is not surprising that this process has dramatic effects on many species of animals and plants.

The application of such principles is not too far from the concept of the fundamental and realized niche, but differs by a more detailed analysis of causes and effects.

Two relatively passive contexts form the border to a more active mosaic in which the decision to stay or to move for animals and to reproduce agamically or by clonation is determined more by the interaction of the specific eco-field with that mosaic.

A comparative analysis conducted using such paradigms can be extremely useful to fill the gap between ecosystem ecology, community ecology, and landscape ecology.

Human Mosaicing

Human use of land creates mosaics, and this seems the most visible human footprint associated with linear elements like roads and railways.

The human mosaic is created by the specialist attitude of man to extract resources, mainly from agriculture. A mosaic is created because land is used differently according to soil climatic factors, soil constraints or for other causes, such as distance from roads, distance from markets, etc.

The human-produced mosaics differ in general from the other types of natural mosaics by the fact that the main cause of the mosaic is not the number of individuals (like in plants and in social animals), but it is the specific use of the natural resources and their handling through agriculture or livestock rearing.

This attitude selects a piece of land with suitable attributes like water content in the soil, geological composition, aspect, distance from natural constraints.

Human mosaicing can be produced according to at least three different strategies:

1. A mosaic is created because resources are patchily distributed and we intercept them according to their spatial arrangement.
2. Or on the other hand the mosaic is created because locally we need several products and we are like central-place foragers. In this way human mosaicing shows patterns common to local human communities. So we can use the land in a patchy way, or we can transform the land in a patchily structured system. It is possible to have a mixing of both strategies. Especially across the Mediterranean basin, such models are common.
3. Finally, we can also use sample resources available in a continuous medium like the cut of a small portion of a forest and then we move to another area in a type of shifting mosaic.

The evolution of the present human mosaic can be observed at different scales. The transformation of a mosaic adapted to central-place foragers to one adapted to net foragers is evident in many regions. Such transformation has modified the spatial arrangement of the mosaics and the diversity of resources available locally. This transformation is a real revolution not only locally but also regionally. In such a way the new mosaics are less diverse locally but they may be more diverse regionally. Products must be transported for longer periods of time and over greater distances because they are delocated from the final consumers. Until the recent past, every local community had a self-organized system to produce goods and services. Every village or small city had surroundings organized in a centripetal way. All around an urban area there were cultivations, and services were provided for the main urban center. Such organization appears very conservative and a real source for local uses and strategies to use resources and to handle goods. Such centripetal organization had a strong impact on the culture, language, and administrative rules. Cultural and socio-economical processes were patchily distributed in the territory contributing to the ecological diversity of a region. In fact the local uses and strategies in resource use have allowed the persistence in a patchy way of species with a distribution greater than the socio-economical patchiness. For instance in some regions a species was strongly menaced by human use of the land, but in other regions a different use of the land allows the persistence of that species.

But the most important consequence of patchy use by humans of the resources has been the creation of an incredible variety of mosaics combining differently patches of woodlands with open field, river margins, roads, edgerow structure and network. Most of the organisms are sensitive to the spatial arrangement of habitat elements that per se are perceived as true habitats (Fig. 4.9).

Today's trend toward a full circulation of goods and services across large regions and continents has disrupted in a short time such systems, reducing the diversity in local land uses and cultural mechanisms. The era of a realized globalization seems very close. The globalization of our time is more broad and sophisticated than the partial globalization during past empires (from Roman times onward). The actual network of roads, the telephone network, and the dissemination of media (radio and

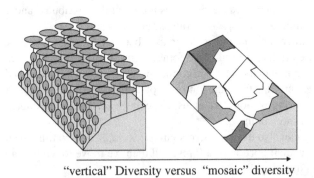

"vertical" Diversity versus "mosaic" diversity

Fig. 4.9 The figure shows the hypothesis that the pre-human environment in the Mediterranean has been substituted by a mosaic of cultivated and marginal patches. The diversity of the pre-settlement environment which we call "vertical" due mainly to the vertical complexity of the primeval forests has been substituted by a horizontal "mosaic" complexity. During this shift, which probably lasted thousands of years, many organisms became extinct while others were favored

TV) via satellites allows a full exchange of information between different parts of humanity and only broad regions still resist this openness. In this process the loss of environmental mosaics reduce the potentiality of ecological diversity to persist.

For instance, across the Mediterranean, despite 13,000 and more years of intense civilization, ecological diversity still persists. Overgrazing, fires, land reclamation, human development, and domestication have reduced but not destroyed the ecological diversity. Botanists have counted thousands of plants living in densely populated areas across the Mediterranean basin (Blondel and Aronson 1999, Rackham 1998, Grove and Rackham 2001), and in many upland areas there is livestock grazing, which allows the persistence of rare species of plants. The use of the land has been so strictly linked with the organism's strategies that often we observe real co-evolutionary processes.

It is reasonable to imagine that the diversity actually observed across the Mediterranean is the result of natural as well as human mosaicing. Such a mosaic has replaced the original more homogeneous land and the more important "vertical" diversity of nondisturbed forest covers present before human settlement (Fig. 4.10).

The mosaic created by human stewardship is not devoted to creating a mosaic per se but to extracting from the land the necessary resources, vegetable and animal biomasses. Type and composition of this mosaic greatly varies according to the agricultural practices. Every crop or any other product requires precise scaled interventions, such as ploughing, fertilization, care for the products, etc.

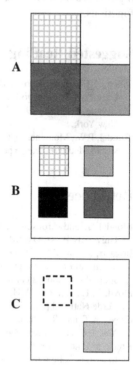

Fig. 4.10 Three different human-induced mechanisms responsible for the formation of mosaics: (**A**) Resources are patchily distributed and human activity copes such spatiality. (**B**) The environmental context is homogeneous and human activity creates an ex-novo patchiness to extract resources. (**C**) The environmental context is homogeneous and human use is temporary (a shifting mosaic)

These activities modify soil composition and dynamics, water cycle and nutrient flux as well as overlapping with the natural patchiness of the land, creating new spatial arrangements that in many cases are recognized as favorable habitats by organisms. In particular grazing by sheep and goats has shaped the mosaic of extensive parts across the Mediterranean. The grazing disturbance influences positively the cover of grass vegetation reducing soil erosion when compared with soil cover in dense shrublands and increases the patchiness of vegetation cover (Perevolotsky and Seligman 1998).

Mosaic Forecasting

The mosaic is the result of different opposing forces or mechanisms, and the study of these structures helps us to better understand ecosystem functions. Forecasting changes of function in the ecosystem through study of the land mosaic is increasingly attracting the attention of scientists. Recently, Bogaert et al. (2002) utilized the shape of mosaic greenness to investigate the behavior of different biomes under global warming. This research is extremely useful for forecasting changes in ecosystem structures around the world.

A mosaic can evolve or it can collapse, both conditions modify the relationships between organisms and their environment.

Evolution in a mosaic means an increase in diversity and complexity of composing patches by internal as well as by external constraints born from self-organizing functions.

Suggested Reading

Allen, T.F.H. and Starr, T.B. 1982. Hierarchy. Perspectives for ecological complexity. The University of Chicago Press, Chicago.

Bissonette, J.A. (ed.) 1997. Wildlife and landscape ecology. Effect of pattern and scale. Springer, New York.

Rundel, P.W., Montenegro, G., and Jaksic, F.M. (eds.) 1998. Landscape disturbance and biodiversity in Mediterranean-type ecosystems. Springer, Berlin.

References

Blondel, J. and Aronson, J. 1999. Biology and wildlife of the Mediterranean region. Oxford University Press, Oxford

Bogaert, J., Zhou, L., Tucker, C.J., Myneni, R.B., and Ceulemans, R. 2002. Evidence for a persistent and extensive greening trend in Eurasia inferred from satellite vegetation index data. Journal of Geophysical Research 107: D11.

Floyd, M.L., Romme, W.H., and Hanna, D.D. 2000. Fire history and vegetation pattern in Mesa Verde National park, Colorado, USA. Ecological Applications 10: 1666–1680.

Grove, A.T. and Rackham, O. 2001. The nature of Mediterranean Europe. An ecological history. Yale University Press, New Haven & London.

Knick, S.T. and Rotenberry, J.T. 2000. Ghosts of habitats past, contribution of landscape changes to current habitat used by shrubland birds. Ecology 81: 220–227.

Kramer, M.G., Hansen, A.J., Taper, M.L., and Kissinger, E.J. 2001. Abiotic controls on long-term windthrow disturbance and temperate rain forest dynamics in southeast Alaska. Ecology 82: 2749–2768.

Laca, E.A., Distel, R.A., Griggs, T.C., and Demment, M.W. 1994. Effects of canopy structure on patch depression by grazers. Ecology 75: 706–716.

Lawrence, R.L. and Ripple, W.J. 2000. Fifteen years of revegetation of Mount St. Helens: A landscape-scale analysis. Ecology 81: 2742–2752.

Levin, S.A. 1976. Population dynamic models in heterogeneous environments. Ann. Rev. Ecol. Syst. 7: 287–310.

O'Neill, R.V., DeAngelis, D.L., Waide, J.B., and Allen, T.F.H. 1986. A hierarchical concept of ecosystems. Princeton University Press, Princeton, NJ.

Perevolotsky, A. and Seligman, N.G. 1998. Role of grazing in Mediterranean rangeland ecosystems. BioScience 48(12): 1007–1017.

Rackham, O. 1998. Implication of historical ecology for conservation. In: Sutherland, W.J. (ed.), Conservation, science and action. Blackwell, Oxford, UK, pp. 152–175.

Scoone, I. 1995. Exploiting heterogeneity: Habitat use by cattle in dryland Zimbabwe. Journal of Arid Environment 29: 221–237.

Steinauer, E.M. and Collins, S.L. 2001. Feedback loops in ecological hierarchies following urine deposition in tallgrass prairie. Ecology 82: 1319–1329.

Summerville, K.S. and Crist, T.O. 2001. Effects of experimental habitat fragmentation on patch use by butterflies and skippers (Lepidoptera). Ecology 82: 1360–1370.

Chapter 5
Ontogenesis and Changes of the Landscape: A Probabilistic View

Introduction

In the second chapter, I distinguished at least three possible approaches to understanding the landscape: the Individual-Based Perceptional Landscape (IBPL), the Individual-Based Cognitive Landscape (IBCL), and the Neutrality-Based Landscape (NBL). These three different perspectives generate processes and patterns that can be perceived moving around. If these perspectives are indicators of real processes and related patterns, we can expect to see some results in a higher-rank meta-domain. The idea that the landscape is a space in which relationships and interactions happen, or is the geography of every domain, is common to all three visions. In other words, the space is the container in which complexity happens continuously. This complexity embarrasses our science which is based more on separate parts than on relationships.

According to Biocomplexity theory (Thompson et al. 2001), landscape is one of the possible dimensions or domains in which the complexity of our planet is realized.

In this chapter, I'll address the problem of the ontogenesis of the "landscape" and deal with the immediately appearing question: what type of landscape do we have to consider?

The aim of this chapter is to develop a theory about the origin of the landscape that helps us to understand the patterns of the present-time mosaics and their evolutionary dynamics. I try to describe possible mechanisms to develop a real landscape in an unknown temporal status without any presumption of discussing evolutionary dynamics and a universal goal function (Wilhelm and Bruggemann 2000).

Several studies deal with the emergent properties of the ecosystems and landscapes, and most of these try to understand the behavior of such systems (see Morowitz 2002 for a synthesis).

For every paradigm we can find empirical evidence, and this seems a general rule in the biocomplexity domain. The infinite complexity of our living and nonliving systems allows us to observe all things that we can imagine.

A. Farina, *Ecology, Cognition and Landscape*, Landscape Series 11,
DOI 10.1007/978-90-481-3138-9_5, © Springer Science+Business Media B.V. 2010

The Functional Circle of the Landscape

As in the theory of meaning, autopoiesis theory, and semiotic science, the concept of circle or closure is adopted to describe phenomena that behave as if they were in a circle. This vision can be useful to describe the ontogenetic mechanisms of landscapes. Considering the three visions of the landscape, the Individual-Based Perceptional Landscape (IBPL), the Individual-Based Cognitive Landscape (IBCL), and the Neutrality-Based Landscape (NBL), the relationships between them are not irreversible and mono-directional but recursively prone. In IBPL, a species needs an existing space in which to select the perceived objects necessary to accomplish functional traits. In IBCL, the observer has the same exigencies described for the IBPL by which to add a conceptual framework to link attributes and to create relationships. At the same time, during the descriptive phase the observer modifies the objects observed. For instance, if we observe a beautiful forest, we could decide to camp inside, producing some types of interference with the natural entities as a consequence of our cognitive evaluative process. Finally, in NBL the IBPL and the IBCL converge at which point we have to add other possible phenomelogical domains that at the present are not observable but forecasted. The circle is completed and I call this the "Landscape ontogenetic closure" in which every component contributes and is part of the other phenomenological domains (Fig. 5.1).

If a landscape is the space of all interactions and relationships, in this space we can distinguish energetic, informative, and semiotic interactions between the different actors that appear in a specific area. I refer to both abiotic actors (physical process) and biotic ones (genetic and semiotic processes).

We try to begin considering the NBL as the more "physical" entity although immediate contradictions appear. In fact, this landscape is the result of involuntary individual-based landscape interactions. In this definition our physical dimension is supported by a cognitive component. Furthermore, in the premises I have also considered the landscape as the result of human observation, or as a creation of our observer perception driven by our culture. Again, the conceptual properties of the landscape are coupled with the physical components.

Whatever we consider, the ontogenetic processes are common to all three visions of the landscape. I use the word ontogenesis specifically to underscore the complex

Fig. 5.1 Representation of the landscape closure created by the three functional-based landscapes (IBPL, individual-based perceptional landscape; IBCL, individual-based cognitive landscape; NBL, neutrality-based landscape)

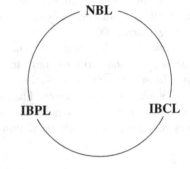

nature of a landscape as a system of several domains that can or can't interact via a meta-domain, or can or can't stay in isolation.

If I consider that every phenomenon is active in a specific domain and that every landscape is the product of forces applied in temporal sequence, I need to introduce the concept of the infinite domain in which space, time, and organization operate. When I describe a present-day landscape, I have to be careful because the organization/time of this system can't be erased. I use the term infinite to stress the huge combinations of ontogenetic conditions that a system can experience.

The Composition of the Landscape

Assuming a neutral landscape that exists outside the individual's explicit perception, in such a landscape energy, information, signals, memory, and cognition are active, all of which are under cybernetic control mechanisms. Forms in a landscape are produced by processes related to the individuals as well as to processes driven by the emergent properties of the multitude of individuals.

Imagine now a completely bare land, without any form of life, and we can then imagine a dynamic that progressively creates life and interactions (see for instance the ecological succession paradigm). Instead of registering the processes and their typology, we register the magnitude and frequency of the distinct processes. At this point we observe a repetition of at least three main processes that I call "opportunities," "events," and "novelties." These three typologies of processes are common to every landscape and for this I consider their description important.

Two main forces are acting in the landscape arena: a subjective force that is represented by the individual-based perception of the landscape [see the eco-field concept (Farina and Belgrano 2004, Farina et al. 2005)] and an "objective" force that is the stochastic result of emergent properties of the units composing the (system) landscape. These two visions operate in meta-domains that do not overlap.

The individual, namely a plant, a bacterium or a man, is an active component of landscape formation changing the neighboring context (the eco-field, Farina 2000, Farina and Belgrano 2004) and contemporarily controlled by components of the same neighboring context. An individual (plant, bacterium, man) or a system contributes to the landscape formation in an active fashion by modifying the neighboring attributes and at the same time is constrained by this context.

When we are dealing with a landscape, the focal point is never a single individual, such as a tree, a bacterium, or a mammal, but the focal point is, at the very least, populations that by aggregation form the patches of a mosaic.

Two main phenomena can be considered as part of landscape ontogenesis: the processes that guarantee the invariance in biological forms (genetic control) and the stochasticity of aggregations of biological forms and associated physical constraints. In this analysis, I'll consider only the phenomena linked to the stochasticity of the aggregation, although it would be an error to underestimate the role of biological form in shaping a landscape. For instance the concept of umbrella and/or key-stone species is very popular in ecology (see Paine 2002).

First we have to assume that the cognitive component of these aggregations (or patches, when distinct from a background) is not explicit like the cognition of individuals. The new emergent properties (created by numerosity) cannot be handled in a direct way because they do not depend on a sequence of codes, as the DNA of individuals, but they are driven by other forces that often escape our capacity of description and finally of understanding.

For these reasons, to deal with the genesis and dynamics of landscapes, a probabilistic language (informative language) is used to identify key processes.

The Probabilistic Model

Considering a landscape composed by a mosaic of units (patches), we can imagine that the ontogenetic processes appear according to a space/temporal caliber and that this caliber produces scaled effects.

We can imagine at least three ontogenetic levels based on a probabilistic typology. These are formally inserted into a hierarchy that is coupled in space and time.

Assuming an observer located away from the systems in a truly neutral position (this is never accomplished) and observing the phenomena that occur inside the system we can note at least three temporal occurrences with distinct frequencies and spatial orientation.

The most frequent and most space restricted occurrences are named "opportunities," and they are paired to the Greek letter α. The second is a stage of the events, β. Lastly there is a stage of the novelties, γ, when the elements external to the system are dominant (Fig. 5.2).

These three states can be copied, maintaining the same sequence, across a broad range of scales. In terms of efficiency in processing energy, material, and organisms,

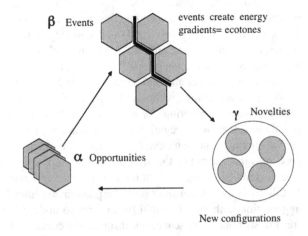

Fig. 5.2 Hierarchical relationship between the three typology of ontogenetic phenomena. Novelties of a lower level create conditions for opportunities at the superior level

it is possible to characterize each state according to a mathematical metaphor: opportunities are like additive operators, events are multiplicative operators, and finally novelties are exponential operators. The energy and the speed of changes in a landscape can be easily understood for each state. The slowest changes occur at the γ level, the fastest at the α level. It is intuitive that at each state different amounts of energy and information are involved.

The frequency of occurrence of these ontogenetic processes decreases moving from opportunities (α) to novelties (γ). Conversely, the memory implicated in the processes decreases moving from novelties to opportunities. Few bits of memory are necessary for active opportunities, but for events and novelties the memory stock plays a central role. The stochasticity increases moving from opportunities to novelties. And finally new landscape configurations are expected in the novelty state.

Opportunities are the most common and frequent state created by the neighboring of different objects. Events are created by the combination of opportunities, but novelties require the contribution of some extra-system energy. The negative entropy level increases from the α state to the β state. This is a signal of increases in system complexity, but this also occurs in moving from β to γ.

According to the amount of energy exchanged between the different levels, the intrasystemic energy prevails at the opportunities state with an increase of coalescence and an increase in interactions.

The event state is dominated by an exchange of intersystemic energy that is concentrated at the ecotones. In the novelty processes energy is supplied from the outside with a dramatic increase in entropy inside the system.

Opportunities can be produced by a disturbance that operates inside the system and determine an accumulation of memory with the possibility of a recursive loop. Also the events are produced by this disturbance process, but their emergent properties are represented by the formation of energetic gradients among the patches that compose the land mosaic and participate in the community's coalescence.

Finally, novelties are created by energy input from the outside that are so strong they create new spatial and energetic configurations. A formidable contribution to new configurations is done continuously by human activities that have the capacity to mobilize large amounts of energy in short time and that focus on sensitive targets.

All the three states have to be considered in a hierarchical framework in which opportunities occupy the lowest level and novelties the highest state.

The α, β, and γ states have a sequence of occurrence that can't be inverted.

Landscape Changes

In the last section I have described three possibilities of landscape ontogenesis. Ontogenesis is a special case of changes in which opportunities are more connected with dynamics, and novelties are by definition more related to the concept of stochastic change.

We summarize briefly the three possibilities by which landscape ontogenesis operates:

"Opportunity" is a process that happens inside an aggregated unit (population, ecotope, or community) and can be considered the first level of a self-organizing process. The energy used to produce changes is shared inside the unit's components.

"Event" is a process that occurs between units (patches) and the allocated energy is extracted by the difference of information that characterizes each patch. The ecotones are the physical "localities" in which events are active. The establishment of ecotones determines changes in species composition and, more generally, an increase in the interpatch connectivity for matter, energy, information, and organisms. It also determines the self-organization of the configuration.

Finally "novelty" happens when a system is supplied by an extra "alien" energy, such as hurricanes in a tropical forest, the invasion of an aggressive species, or an earthquake. In this case, a new configuration appears in the system, substantially changing the history of the system and the direction of its dynamics.

According to the dissipative nature of living systems (Prigogine 1961, Odum 1983) and the opposite self-organization attitude of complex systems (Kauffman 1993), landscapes are shaped and modified continuously by internal and external forces that create order (information, sensu Stonier 1990, 1996) followed by a disordered, entropic state that shows oscillating dynamics (Holling 1973).

Landscapes can be considered dense, complex, networks of interactions between environmental factors (both forms and patterns, processes, and dynamics) that capture implicit as well as explicit properties of the environment (Phillips 1999).

But in order to maintain and sustain such a complex aggregate system (sensu Manson 2001), changes seem to be fundamental processes common to every scaled level of resolution at which a system is investigated.

The Dual Nature of Complex Systems

Several definitions are available for complex systems (see Lewin 1999, Cilliers 1998). Teleonomy, autopoiesis, and self-organization seem to be key concepts to understand the nature of a complex system.

Complexity can be created and maintained by a teleonomic (project) mechanism (Monod 1970) at the level of a biological entity, and by a goal function at the level of a natural system (Wilhelm and Bruggemann 2000).

In the next section I will discuss the changes that occur in teleonomic entities as well as in emergent entities in which goal functions can be described by emergent properties.

The Nature of the Changes

River meandering, oscillation in the geographical ranges of species, land abandonment, forest fragmentation, reclamation, and urban sprawl are some of the changes that commonly occur in natural and modified environments.

A change can be defined as any modification occurring in a system state (from individual to biosphere) produced by a broad variety of abiotic and/or biotic factors that introduce or subtract energy and information to the system.

Changes can be considered modifications in the availability of an expected resource or pattern and the temporary or permanent impossibility for species, populations, communities, ecosystems, and land mosaics to incorporate the new conditions.

Every type of change is scaled differently, as are the processes involved. The processes responsible can operate at the individual-based landscape on the emergent properties at the neutrality-based landscape scale.

Landscapes, which are also considered interacting collections of patches, are continuously under change, and it is possible to distinguish changes inside the components and changes that occur when external forces act on the system. But a landscape can also be considered a container of different levels of organized complexity according to a hierarchical perspective (Allen and Starr 1982, Allen and Hoekstra 1992) and changes occurring at one level can be integrated with changes at other levels.

Changes in the environment are "relativistic" properties. Without a standard of comparison, we cannot evaluate a change. Changes occur in space and time separately or contemporarily. Changes occur at individual, population, community, ecosystem, and landscape levels.

We will focus on the individual level as well as on the landscape level, considering the population and community levels as "secondary" and "paradigmatic" levels of resolution and not the primary front of "genuine changes." However, emergent properties of populations and communities do have relevant effects on the scaled dynamics.

The "Human" Perception of Changes

Changes occur at every level of the ecological organization from individuals to the entire biosphere, but also at every time resolution. This is one of the main characters of the complexity of the ecological systems (see Manson 2001).

The life span of landscape patterns as perceived by humans (in space and time) is considered an important metric to evaluate the rate of changes that have occurred in the spatial configuration of land mosaics.

Memory of changes can be stored in several devices by man, and this can help us to change the short-term capacity of man to perceive changes. When you observe the picture of your driving license, your comment is always "My God, how young I was!" But this is not appreciated when every morning we see our face in the mirror! This means that changes can be evaluated only if the comparison can be done on a time interval sufficient to appreciate the differences. We perceive the turnover of land spatial configurations because we can resume the history and memory of land mosaics (e.g. using maps, airplane, or satellite images) and because we have the tools (e.g. GIS), to compare the different configurations.

Changes and Dynamics

In the usual sense, changes are processes that modify the structure and/or the dynamic of a system. But the dynamic is at the same time a type of change. Probably a way to clarify this concept is to distinguish a dynamic, as a product of a self-organizing system, from a sudden unpredictable event that produces changes. The history of the system is not the main cause of changes "sensu stricto," but only the reference system. Historically, dynamics have played the main role, determining the sequences of species, pattern, or process turnover. Succession in plant communities is a well-known example of dynamics that develop in a historical context. In other words, dynamics refer to changes linked to a sequence of nonrandom events. In this case, we could call this a patterned history as opposed to a stochastic history in the case of an unpredictable disturbance.

Changes generally are considered in terms of space and time and rarely in terms of function, but see Tilman for the concept of ecological debt (Tilman et al. 1994).

Functional changes capture both the emergent properties of a system as well as the functionality of individuals.

Suggested Reading

Hansson, L., Fahrig, L., and Merriam, G. 1995. Mosaic landscapes and ecological processes. Chapman & Hall, London.
May, R. and McLean, A. 2007. Theoretical ecology. Oxford University Press, Oxford.
Morowitz, H.J. 2002. The emergence of everything. Oxford University Press, Oxford.
Odum, H.T. 1983. System ecology. John Wiley & Sons, New York.

References

Allen, T.F.H. and Hoekstra, T.W. 1992. Toward a unified theory. Columbia University Press, New York.
Allen, T.F.H. and Starr, T.B. 1982. Hierarchy perspectives for ecological complexity. The University of Chicago Press, Chicago.
Cilliers, P. 1998. Complexity & postmodernism. Understanding complex systems. Routledge, London.
Farina, A. 2000. Landscape ecology in action. Kluwer Academic Publisher, Dordrecht, the Netherlands.
Farina, A. and Belgrano, A. 2004. The eco-field: A new paradigm for landscape ecology. Ecological Research 19:107–110.
Farina, A., Bogaert, J., and Schipani, I. 2005. Cognitive landscape and information: New perspectives to investigate the ecological complexity. BioSystem 79:235–240.
Holling, C.S. 1973. Resilience and stability of ecological systems. Annual Review of Ecology and Systematics 4:1–24.
Kauffman, S.A. 1993. The origins of the order: Self-organization and selection in evolution. Oxford University Press, New York.
Lewin, R. 1999. Complexity. Life at the edge of chaos. The University of Chicago Press, Chicago.
Manson, S.M. 2001. Simplifying complexity: A review of complexity theory. Geoforum 32: 405–414.
Monod, J. 1970. Il caso e la necessità. Mondadori, Milano.

Morowitz, H.J. 2002. The emergence of everything. Oxford University Press, Oxford.

Odum, H.T. 1983. System ecology. John Wiley & Sons, New York.

Paine, R.T. 2002. A conversation on refining the concept of keystone species. Conservation Biology 9: 962–964.

Phillips, J.D. 1999. Divergence, convergence, and self-organization in landscapes. Annals of the Association of American Geographers 89(3): 466–488.

Prigogine, I. 1961. Introduction to thermodynamics of irreversible processes. John Wiley, New York.

Stonier, T. 1990. Information and the internal structure of the universe. An exploration into information physics. Springer-Verlag, Berlin.

Stonier, T. 1996. Information as a basic property of the universe. BioSystems 38: 135–140.

Thompson, J.N., Reichman, O.J., Morin, P.J., Polis, G.A., Power, M.E., Sterner, R.W., Couch, C.A., Gough, L., Holt, R., Hooper, D.U., Keesing, F., Lovell, C.R., Milne, B.T., Moles, M.C., Roberts, D.W., and Strauss, S.Y. 2001. Frontiers of ecology. Bioscience 51: 15–24.

Tilman, D., May, R.M., Lehman, C.L., and Nowak, M.A. 1994. Habitat destruction and the extinction debt. Nature 371:65–66.

Wilhelm, T. and Bruggemann, R. 2000. Goal function for the development of natural systems. Ecological Modeling 132: 231–246.

Chapter 6
The Ecotones

Introduction

What happens when a patch enters into contact with another patch? This condition is the rule and not the exception in a mosaic. The distinctive characteristic of a patch is in contact with other characteristics of the neighboring patches and it is possible that it receives a negative reaction (for instance a different pH of soil in the case of a patch of vegetation). On the other side the patch encounters a favorable condition and expands damaging the neighboring patch. For instance, a clone of bamboo spreading into grassland. This case is typical for invader plants. An intermediate reaction is also possible.

Borders of any type are critical areas, either for the physical characteristics of objects or for biological and ecological entities.

Organisms have created barriers at the border of their bodies such as cellular membranes or specialized tissue (tegument). Borders also play an important role in informing the organism that it is at the limits or margin. For this reason, a border (or ecotone), is a very informative area of an individual or of a patch (Fig. 6.1).

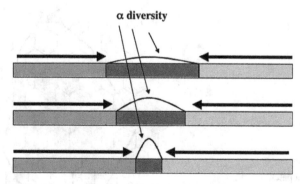

Fig. 6.1 Ecotones can be created by contraction of a patch due to fragmentation or substitution with another patch type. In this case we have considered the case of a woodland that has been progressively reduced by cultivations to a narrow band of natural vegetation assuming the role of an edgerow. The diversity along an edgerow increases due to the margin effect

A. Farina, *Ecology, Cognition and Landscape*, Landscape Series 11, DOI 10.1007/978-90-481-3138-9_6, © Springer Science+Business Media B.V. 2010

The information available to an individual (plant, animal, microbe, fungi, virus) is quite different compared to the information that is collected by a patch (population, community), but in the end the result seems the same. The limits of a favorable area or the possibility for a new area of expansion are detected by sensors in individuals and by the different behavior of components in the population and community patches.

Ecotones have received great attention from ecologists for over 100 years, especially in the context of transition between biomes, geographic vegetation unit, movement of tree lines, and wildlife habitat (Fig. 6.2).

Since 1905, Clements used the term "ecotone" from the combination of two words (eco) oikos (home) and tonos (tension). The American naturalist Leopold (1933) described the greater richness of wildlife at the edges (across the ecotone).

Ecotones contribute to the complexity of the environmental mosaic as well as to the local complexity, either in terrestrial or in fresh and salt water bodies.

It is not easy to study ecotones because of their fuzzy character when observed at a fine scale, and their temporal and spatial instability increases such difficulties.

Ecotones can be observed across many scales, and we can consider the ecotones inside the paradigm of the hierarchy (Gosz 1993, Wiens 1992).

Initially these were considered as static entities like new habitats, but later, and especially during the past two decades, they were considered edges of a dynamic mosaic. Ecotones have been recognized as important structures regulating the flux of nutrients, alerting to the border of the individual species habitat. They contain high levels of biological diversity in addition to primary and secondary productivity.

Their importance in the functioning of species and also of ecosystems and landscapes has been largely recognized (Risser 1995).

Ecotones are considered interruptions of physical or biological structures along a gradient (Fig. 6.3).

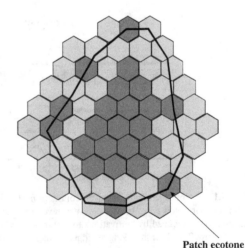

Fig. 6.2 When a patch (population, community) meets another patch the transition zone is characterized by uncertainty; an ecotone is located there

Patch ecotone

Fig. 6.3 Ecotone properties largely depend on their environmental context (**A**). The direction of the dominant wind or the gravitational force can produce severe changes in ecotone function. For instance in image (**B**) the gravitational effects are depicted as the main constraint for the fluxes of matter and nutrients

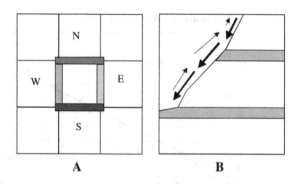

A   B

According to the theory of the mosaic, ecotones represent the margin of expansion or reduction of every patch, the transitional point between different conditions.

We can locate ecotones everywhere in a landscape according to the type of mosaic perceived. So also for ecotones we have to adopt the paradigm of a subjective mosaic strictly linked to the cognitive component of each living organism.

Ecotones exist on all scales, from the biome to a few centimeters in space and from the long term of thousands of years to the ephemeral life of temporary ponds.

Types of Ecotones

We can describe two types of ecotonal effects: a process effect and a pattern effect. The first is the result of a peculiar environmental constraint on the life trait of an organism (plant, animal, virus, fungus, bacterium) as intercepted by the species-specific eco-field. The pattern effect is the result of neighboring of a population or community patch at the margin of their existence (Fig. 6.4a, 6.4b).

Pattern effects are considered in a dynamic process of movement of the limit of such ecotones; for instance, the change of the tree line due to climatic changes or the substitution of one vegetation patch with another. This process may be gradual, abrupt, or intermediate. The graduality is quite common in community patches while abrupt changes occur especially in population patches.

The Spatial Arrangement of Ecotones

Generally, we are used to evaluating an ecotone according to the level of spatial contrast recognized between two or more patches. The level of ecotonality, i.e. the net effect that such a border produces on a species, depends on many factors.

We have ecotones between individuals, populations, and communities. For each entity the ecotones behave differently.

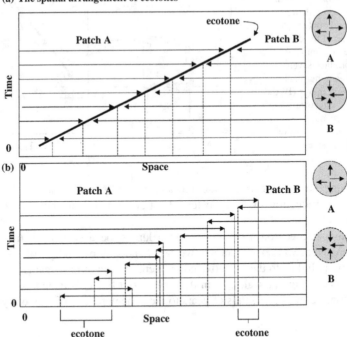

(a) The spatial arrangement of ecotones

Fig. 6.4 The ecotone changes position in space and time due to internal as well as external dynamics. (**a**) In this case patch A substitutes patch B by progressive exclusion. (**b**) In this case patch A substitutes patch B in an irregular way and the ecotone varies in width according to the time sequences

Another relevant difference affecting ecotone characteristics is the position of the borders according to environmental gradients. For instance, a community patch placed in the inner part of a slope can receive more nutrients than patches placed at the top.

The Nature of Ecotones

We can use many definitions for ecotone concepts, but we prefer to consider an ecotone more as a paradigm than a structure. In fact only if we recognize the heterogeneous and patchy nature of a landscape/habitat/ecosystem, can we utilize the ecotone as a proto-principle.

The ecotone may be considered according to different points of view as:

the border between two patches,
the border between two levels of dynamism,
the border between two different states of biological complexity.

Ecotones are everywhere and are indeed the internal as well as the borders of the ecological mosaics! But internal borders generally have lower potential information to transfer between patches than external borders that surround a mosaic distinguishing unique emerging properties (Fig. 6.4).

Such a border in general has a difference of potential like a cellular membrane. The ecotones are the physical location in which a mosaic can be recognized and in which many functions of complex systems are working.

Ecotones have a high level of information, and their predictability is very low. The amount of ecotones in a matrix produces significant effects on the functioning of the entire mosaic in which they are embedded.

The ecotone paradigm is strictly linked to shape and size of fragments and to the effects of such patterns on the functions of many organisms.

Matrix and Ecotones

If we use the paradigm of the ecological matrix to analyze the environmental complexity we can consider the ecotones as the projection on the surface of the elements deeply anchored in the matrix bottom. But again the origin of such functional structures is determined by the interference between matrix processes and species-specific processes.

Matrix complexity has the capacity to produce emergent processes that in turn create the discontinuity that we can observe at the surface.

The Origin of the Ecotone (Effect)

Every border has different characteristics in respect of the central parts of a distribution. The defensive role or the offensive role of border behavior is a general rule for individual entities as well as for populations and communities. According to this universal paradigm, we can describe ecotone properties according to the different entity considered.

Moving from individual to community, ecotonal properties have in common the variance in the explicit phenomena.

Under the different environmental conditions in a population, the allelic frequency of a gene can change creating a genotone (Dobzhansky et al. 1977). An individual intercepts a discontinuity changing the behavior. A population close to an ecotone can change the density of individuals, and a community the composition of species.

A population can increase or decrease the density according to the presence of a less favorable (decrease in local density) or more favorable (increase in local density) habitat (Fig. 6.5). A community can have a richer or poorer composition at the border accordingly.

Fig. 6.5 Change in local density of a population at an ecotone. In **A** the border represents a more hostile field and the population shows a rarefaction. In **B** the border is an attractive place and individuals are more concentrated around it

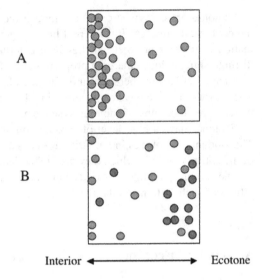

Interior ◄─────────────► Ecotone

Shape of Ecotones

The morphology of the borders is of great importance for the functioning of the ecotones and for the role that they can play in the dynamics of a mosaic. Several empirical studies have demonstrated that the more a border is convoluted, the more complex are the processes that can be observed (Fig. 6.6). Many species of animals

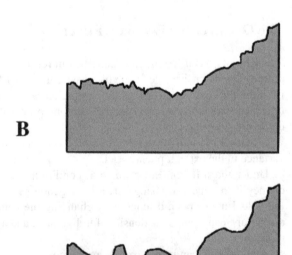

Fig. 6.6 The border of a patch may be highly convoluted as in (**A**) or less convoluted as in (**B**). Generally human activity tries to reduce border irregularity depressing the chances of species to allocate resources

find a convoluted margin more attractive than a rectilinear one. This is well known by wildlifers and environmental managers that try to increase the complexity of the borders to attract more fauna. In many regions where human intrusion has been massive the only seminatural spaces in which spontaneous vegetation and associated biodiversity can exist are between croplands. In such a case, ecotones play a multirole as habitat per se, as corridors, and as a tension zone between two or more different entities.

The ecotones may be of fuzzy or of abrupt type. Every patch tries to isolate itself from the neighboring patch.

We expect that across tension zones the fluxes of energy, material and animal, are at higher levels compared with the patches that compose the mosaic.

Ecotones are created by several mechanisms such as plant succession, by the disturbance regime of grazing animals, and by the direct perception of a functional discontinuity. Some ecotones are created by soil discontinuities affecting the vegetation. For instance, a chalky soil and a siliceous soil support different plant communities, and at the edge between the two soils we can expect an ecotone. Climate can create an ecotone either at micro-climatic scale (altitudinal gradients along a mountain slope) or at macro- or mega-climatic scale (e.g. the Mediterranean and Continental areas) (Fig. 6.7).

Fig. 6.7 Example of a mosaic of upland vegetation composed of meadows, pastures, and scrublands. Ecotones are at the border of each patch. We can also consider the scrubland as an ecotone between meadows and pasture

Changes in topographic, edaphic, or climatic factors create inherent *edges*, but when the edge is the product of external causes like fire, deforestation, or grazing these edges are called induced edges.

The length of an ecotone may be extremely variable according to the typology. Some ecotones have an ephemeral life, like the snow melting line, the gradient of humidity around a pond, etc.

According to their position in space, we can divide ecotones into two broad categories:

The horizontal ecotones have extension along planar surfaces.

The vertical ecotones are localized especially in the atmosphere and in water bodies.

Human use of the land increases the frequency of ecotones that generally are narrow, with abrupt margins and less complex than the ecotones created by natural processes.

The main structural characteristics related to the physical composition are: size, shape, biological structure, structural constraint, inner heterogeneity, ecotonal density, fractal dimension of the edges, diversity of patches, and patch dimension (Hansen et al. 1992).

The functional character pertains to the persistence, resilience, functional constraint, and porosity.

Persistency is the capacity to preserve function and structure under the pressure of a modifier.

Resilience is the capacity of an ecotone to maintain the original characteristics after a disturbance.

The functional constraint is the result of the difference between composing patches.

Porosity is the capacity of an ecotone to modify the flux of energy, material, and organisms. These characteristics have to be calibrated according to the context in which they are expressed (Wiens et al. 1985).

The ecotone paradigm is very useful to explain the dynamics among patches. Inside a patch the centripetal processes are dominant, while the centrifugal process is prevailing at the border.

Inside a patch, we can imagine a spiral that is attracting energy and material from outside; the tension zone is represented by the ecotone.

Ecotones exist only as a function of a species that detects an environmental discontinuity: the properties of an ecotone are strictly linked with the character of the species. The same discontinuity can be trespassed easily by one species while another species may find it difficult or impossible (Fig. 6.8).

According to the characteristics of a species, we can generalize as follows: a mobile species intercepts the ecotones in a softer way compared with a more sedentary species.

According to the living trait, we can describe at least four types of species: species that need a mosaic of patches and then must have ecotones in their home-range, species that live only inside the ecotones, species that prefer only a certain

Fig. 6.8 Five possible strategies for species to interact with an ecotone: *a*: neutral, *b*: diffraction, *c*: reflection across the furthest border between (**A**) and (**B**), *d*: reflection at the nearest border between (**A**) and (**B**), *e*: living in an ecotone (ecotone = habitat)

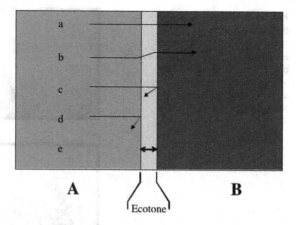

type of habitat but that can cross the ecotone easily, and finally species that consider ecotones as true barriers.

Ecotones generally are created by plants but also large flocks of animals like gazelles can also create ecotonal effects on individuals that are at the periphery. These individuals are more exposed to predation but are the necessary buffer for the entire flock survivorship. Although living at the edge can have some advantages relating to reduction of intraspecific competition, they pay a very high price due to predation.

Generally ecotones are composed of two neighboring patches, but in some cases more than two patches meet, and in this case this area is called "covert" and is particularly dynamic.

Moving across a spatial scale, we can find ecotones from the fine scale of a few meters to the large scale of a river catchment. It seems that enlarging the spatial scale of the ecotones increases their importance for the functioning of the entire system. For instance, the rivers can be considered very important ecotones along which the high dynamicity of water and the diluted substances move organisms and energy.

Population distribution is characterized by real borders and edges. The processes that occur at the margins are important to understand: limit of ranges, responses to environmental change, genetic divergences, and speciation.

Spatially explicit models can help to understand the behavior of populations at the margin, considering the ecological as well as the genetic constraints.

The behavior of a population at the edge may simply be diffusive in which adults are mobile but no births occur (Fig. 6.9) or by a more realistic model in which there are no movements of adults but dispersal of newly born individuals (see f.i. Antonovics et al. 2001).

Cumulative distribution has more complex margins than real-time populations. This depends on the cumulative effect of time and on the fact that rare events are considered important like common ones. The marginal populations can have an important role as refugia from invading pathogens, like the American chestnuts that are resistant to chestnut blight at their margins. At the margin individuals are more

Fig. 6.9 Distribution of
population margin according
to a diffuse model: (**A**) adults
are moving but there are no
births, and (**B**) no movements
of adults but dispersal of
newly born individuals (from
Antonovics et al. 2001)

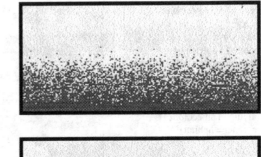

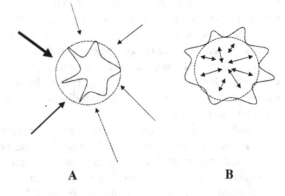

isolated and distributed in a more unpredictable way having scanty contact with the
core population.

Patch formation is strongly connected with two fundamental processes: the first
is the environmental constraint and the second is the internal (genetic constraints).
Both are not independent but are linked by feedbacks (Fig. 6.10). The limit of patch
populations is often very sharp and not coincidental with evident soil discontinuity
(for plants) or climatic stress. According to Antonovics et al. (2001), some causes
of plant patchiness could be found in the genetic properties of the species, especially in the gene flow that reduces the potential adaptability to new conditions if
the source of this flow is from the same population locally differentiated for a long
time. The growth of the population depends on continuous immigration from the
source population, on gene flow (seed and pollen) that can limit the adaptation of

Fig. 6.10 Patch ontogenesis
may depend on external
constraints like landslides,
wind storms, or lava flows
(**A**), or on internal dynamics
like intraspecific competition,
and gene flow that prevents
individual adaptation (**B**).
Often both processes are in
action

pioneering plants, and finally on stochastic processes where deterministic processes are impossible.

Also if the surroundings are favorable for that species, often ecological and genetic factors can inhibit the expansion of a population that remains under a constant size and shape as if embedded in a hostile matrix.

Often populations at the margin have a "flame-like" behavior, ephemeral in time, and the cumulative occurrence of species can be more fractal than the actual distribution. Pathogens at the margin of a population are less efficient in attacking a sparse population. The gene flow from a central distribution of a population depresses the adaptability of pioneering individuals, and also sparse and isolated individuals can suffer from genetic drift and accumulate deleterious mutations. Competition can play an important role and can maintain sharp borders between species. The limited number of individuals at the border of a population can increase the gene flow influence producing failure in adaptation capabilities.

Suggested Reading

Holland, M.M., Risser, P.G., and Naiman, R.J. (eds.) 1991. Ecotones. The role of landscape boundaries in the management and restoration of changing environments. Chapman & Hall, New York.

Hansen, A.J. and di Castri, F. (eds.) 1992. Landscape boundaries. Consequences for biotic diversity and ecological flows. Springer-Verlag, New York.

Malanson, G.P. 1993. Riparian landscapes. Cambridge University Press, Cambridge.

References

Antonovics, J., Newman, T.J., and Best, B.J. 2001. Spatially explicit studies on the ecology and genetics of population margins. In: Silvertown, J. and Antonovics, J. (eds.), Integrating ecology and evolution in a spatial context. The 14th Special Symposium of the British Ecological Society held at Royal Holloway College, University of London, 29–31 August, 2000. Blackwell Science, Oxford, pp. 97–116.

Dobzhansky, T., Ayala, F.J., Stebbens, G.L., and Valentine, J.W. 1977. Evolution. Freeman, San Francisco.

Gosz, J.R. 1993. Ecotone hierarchies. Ecological Applications 3: 369–376.

Hansen, A.J., di Castri, F., and Naiman, R.J. 1992. Ecotones: What and why? In: Hansen, A.J. and di Castri, F. (eds.), Landscape boundaries. Consequences for biotic diversity and ecological flows. Springer-Verlag, New York.

Leopold, A. 1933. Game management. Scriber, New York.

Risser, P.G. 1995. The status of the science examining ecotones. BioScience 45: 318–325.

Wiens, J.A. 1992. Ecological flow across landscape boundaries: A conceptual overview. In: Hansen, A.J. and di Castri, F. (eds.), Landscape boundaries. Consequences for biotic diversity and ecological flows. Springer-Verlag, New York, pp. 217–235.

Wiens, J.A., Crawford, C.S., and Gosz, R. 1985. Boundary dynamics: A conceptual framework for studying landscape ecosystems. Oikos 45: 421–427.

Chapter 7
Measuring and Evaluating the Ecological Mosaics: General Assumption

Introduction

Mosaics have common properties that allow us to apply well-tested indexes to measure their spatial properties.

Size, shape, and context are the more important attributes of each mosaic. The meaning of such measures that are often used without a clear statement is fundamental. Some measures are redundant or statistically correlated.

Nevertheless, today many indexes are available to evaluate a mosaic: Turnover, contagion, lacunarity, diversity, dominance, fractal characters are some of the more used metrics.

Most of these indexes are derived from landscape ecology, others from community or population ecology, but all are extremely important for evaluating the complexity of the mosaics.

The indexes can be divided into two broad categories: structural indexes and context indexes.

The structural indexes measure the inherent characteristics of a patch:

- Size (dimension of largest, smallest, etc.)
- Shape (relationship area/perimeter)
- Patch-matrix contrast
- Turnover (temporal replacement of patch types)
- Contagion (level of aggregation)
- Lacunarity (level of dispersion)
- Diversity (richness)
- Evenness (equal distribution)
- Dominance (level of concentration)

To evaluate species in a mosaic we can use spatial statistics like spatial autocorrelation defined by Sokal and Thomson (1987) as "the dependence of the value of a variable on values of the same variable at geographically adjoining locations." See also Hu and Moskat (1994).

A. Farina, *Ecology, Cognition and Landscape*, Landscape Series 11, DOI 10.1007/978-90-481-3138-9_7, © Springer Science+Business Media B.V. 2010

Generally the structural indexes deal with size and shape of the composing patches of a mosaic.

The context indexes measure the spatial attributes of the different categories of patches and the overall characteristics of the mosaic per se.

All the indexes can be applied efficiently only if a comparative action follows the measures.

The structural indexes are based on the assumption that every component of a mosaic is the product of interfering processes and that every typology of patches has per se a different behavior and role in the environment. Every patch may have specific characters, but it is not always true that these characters are perceived by a species that utilizes the patch. It seems important to calibrate the analysis to the organisms which we are interested in investigating. Often we use maps created by our interpretation of the mosaic and this is the great limit of spatial analysis.

A mosaic fragmented for a species can only be heterogeneous for another species (Fig. 7.1). The term fragmented means insulation of suitable patches, dispersed into a hostile matrix. Heterogeneous means that the resources are not found everywhere inside the suitable matrix.

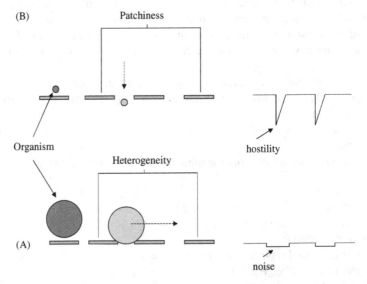

Fig. 7.1 The different perception of the environmental discontinuity according to the specific caliber. A species can intercept an environmental discontinuity as a noise (heterogeneity) (A) or the patchiness as a barrier (hostility) (B)

The Patch Size

The dimension of patches has inherent related consequences. A small patch means a population of small dimension with greater external influence reaching the inner parts.

We know that small animals have higher exchanges with the external medium. For instance, a hummingbird or a firecrest (*Regulus* sp.) has to spend more energy to maintain internal homeostasis than an elephant. The same mechanism is active in small patches that become sensitive to external influence.

For instance, small patches of shrubs are more easily burned than large ones because of the surrounding invasive highly inflammable grasses. So the persistence of small patches is less certain than large ones. In some cases, small patches can survive better than large ones if a disturbance regime is acting against it. For instance, along a tree fall on steep mountains small patches of vegetation have less probability to be intercepted by rock outcrop, but this is only an example of a specific situation; the opposite is the rule. From an organismic perspective a small patch may not be enough to host a species, or only a limited number of individuals can be supported. Again in this case a reduced population of hosts can be supported, or only a few parts of a habitat patch can be considered source.

Some species are particularly sensitive to the border of suitable habitats and for this reason need a large extension of "habitat" far from this border. These species have been called "interior" species. The internal part of this habitat is called the "core area." In theory, two hypotheses can be presented: The first is that at the border of a habitat there is no "habitat" for that species, and the second hypothesis is that a less suitable habitat is placed at the periphery and that this habitat is populated only by sink populations that survive only by the refueling of individuals lost by new arrivals from source habitats.

The Patch Shape

The shape of patches assumes a relevant importance, for the maintenance of the patch per se and as recognized habitat for a focal species.

Highly fractal patches offer more surface of contact with other patch types, and, under conditions of competition or of dominance of the neighboring patches, the shape of the border can encourage the subordinate patch to replace the invisible patch.

The fractal nature of the border can be estimated simply as the ratio between the total length of the border and internal surface of the patch. Or better still, it is possible to measure the convolution of the border applying a fractal analysis, for instance using the mass block.

The length of a patch border is not an inherent characteristic of a patch, but depends on the length of the caliber employed to measure this length.

For this reason we can measure the length of a patch using the meter to characterize the patch, but in many cases this measure is out of the range of appreciation of a species. There are good examples in the literature of employment of the species caliber to assess the length of a border. Using this method, it is possible to include also the irregularities of a border.

The convolution of the border can be perceived as important for some species like deer when they are foraging at the border between woodland and open grassland. But again the great extension of the borders facilitates the presence of edge species, of predators, and de facto the great extension of the border increases the sink character of the populations living in this system.

The Patch-Matrix Contrast

This is an index that is quite difficult to calculate. Especially in a human disturbance environment, we can utilize such parameters to better understand the rate of energy, material, and organism exchange between two neighboring patches or between a patch and the surrounding matrix.

For instance, between a woodlot and an open cropland the contrast between the two typologies is very high. Along a successional gradient this could be quite soft allowing a transaction from the two typologies.

How does one measure this contrast? Generally the height of the vegetation and/or the density (stem $\times$ m^2, etc.) are employed.

The Turnover

The patch turnover simply means the expected or observed time of replacement of a patch with another patch, or the recovery of the same type of patch after a disturbance event (Fig. 7.2).

Fig. 7.2 A patch can be replaced by another patch according to disturbance or succession

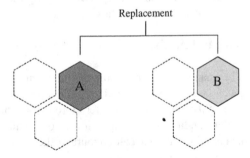

The mosaic-cycle concept has been extensively used by vegetationists to explain the movement of vegetation according to patchy patterns.

Especially for plants with a very rapid growth rate, spatial turnover is the rule. For instance, annual plants can appear suddenly, bring pattern to the system and then as quickly as they sprouted, they may disappear.

The patch turnover can be produced by disturbances, such as flooding, wild fires, wind storms, tree fall, animal grazing, trampling, digging, or burrowing.

The Contagion

Contagion measures the level of aggregation of a patch system compared with other types of patches. This index is sensitive to the aggregation of elements inside a patch. For instance, the maximal level of contagion in a matrix composed of two

Fig. 7.3 In matrix A the two covers have the maximum possible contagion for that configuration. In matrix B the contagion has the minimum possible for that configuration and the patches appear interspersed

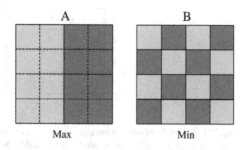

different types of patches is observable when the two typologies each occupy half of the matrix. The minimum is reached when the two covers are distributed like a checkerboard (Fig. 7.3).

The Lacunarity

The lacunarity is a measure of spatial arrangement between two or more patches. This index focuses on the empty spaces between patches and is particularly sensitive to their spatial arrangement. In particular lacunarity is important for processes or species patterns that happen outside the habitat patches, like dispersion, predation, and settlement. Lacunarity measures the pioneering capacity of species, or the environmental pressure of the more or less hostile environment on a species that is in movement from one habitat patch to another.

The Diversity Index

Diversity or richness is one of the most used and abused indexes in ecology. The precise meaning of diversity has its roots in information theory, in the probability of finding in a sampled collection of objects new objects at every step.

The diversity index applied to the mosaic describes the typologies of the patches, their variety according to some pre-established patterns.

When we measure the diversity of a vegetation mosaic, we can use categories like old growth forest, regeneration forest, secondary forest, logging, etc. These characters are selected according to our purposes and must be calibrated according to the focal organisms or process under investigation. In some cases the diversity is simply the number of pixel types inside an area determined for our purposes.

Diversity in remote sensing is used to evaluate the complexity of the neighboring area and to relate such complexity (in this case equal to diversity) to processes like presence-absence of species, abundance of a focal species, and distance from diverse sites.

a	b	g	h
c	d	m	n
e	f	o	r
i	l	q	p

A

a	b	a	h
a	d	a	a
a	a	a	a
a	l	a	p

B

Fig. 7.4 Example of distribution of patch types in two matrixes. Matrix A is characterized by the maximal diversity (H′) and evenness (J′) possible. Matrix B has a low diversity and also a very lower evenness. Where $H' = -\Sigma\ p_i \log p_i$, $J = H'/H_{max}$, $H_{max} = \log S$, S = Number of categories

The use of this index in remote sensing is particularly useful, especially when the information on the land is partial and when for instance, multispectral images are not available (Fig. 7.4).

Diversity of pixels is strictly connected with the scale of observation, namely the amount of area in which the diversity has been computerized.

The Evenness

This index calculated after the diversity indicates how the richness of a collection is concentrated in a few species. Size and number of each patch type are considered important descriptors of a mosaic. The closer the value of evenness is to 1, the more the patch types are equally distributed between categories (Fig. 7.4).

The Dominance

This index measures the level of concentration of the patches and is the opposite of evenness. Dominance can be calculated as a ratio between 1 and the diversity.

Suggested Reading

Burrough, P.A. and McDonnell, R.A. 1998. Principles of geographical information systems. Oxford University Press, Oxford.
Gergel, S.E. and Turner, M.J. 2002. Learning landscape ecology. Springer, New York.
Klopatek, J.M. and Gardner, R.H. (eds.) 1999. Landscape ecological analysis. Issues and applications. Springer, New York.
Turner, M.G. and Gardner, R.H. (eds.) 1991. Quantitative methods in landscape ecology. Springer-Verlag, New York.

References

Hu, Z.J. and Moskat, C. 1994. Effect of habitat selection strategy on spatial correlograms in a heterogeneous environment: A simulation study. Acta Zoologica Academiae Scientiarum Hungaricae 40: 369–377.

Sokal, R.R. and Thomson, J.D. 1987. Applications of spatial autocorrelation in ecology. In: Legendre, P. and Legendre, L. (eds.), Development in numerical ecology, NATO ANSI Series, Vol. G14. Springer-Verlag, Berlin, pp. 431–466.

Chapter 8
The Cognitive Landscape

Introduction

Scientific thought as we know it today, is based upon the assumption of an objective, external world. This conviction is supported by a rationale that calls upon mechanical laws of causal efficacy and determinism. Fundamentally, the correspondence between the hypotheses and their predictions through experimental research build the empirical success of Science. Nature, however, does not conform to the conditions required by a classical thermodynamic theory in which a physical state irreversibly evolves from its most probable, uniform, inert and unchangeable macroscopical precursor. The reductionist approach applied to ecology has shown limitations within its own domains of excellence. The homeostatic balances directed by determinism are often accompanied by unforeseeable instabilities where the dissipation of energy becomes the source of new macroscopic structures (May 1974, 1976, 1986, Kauffman 1993, Prigogine and Stengers 1984).

There is presently considerable interest in understanding the complexity of ecosystems (Merry 1995, Cilliers 1998, Bradbury et al. 2000, Manson 2001). Many disciplines have an interest and a contribution to make, but barriers between disciplines are a major limiting factor for the development of an integrated problem-solving science (Graham and Dayton 2002) with the capacity for a higher ethical profile (Haber 2002). However, the separation between social, economic, and environmental sciences is a matter of fact, and progress in developing suitable interdisciplinarity appears to be a requirement for the advancement of Science. A rich theoretical framework developed during the past few decades (General System Theory by von Bertalanffy 1969, autopoietic organization by Maturana and Varela 1980, zoosemiotic by Sebeok 1968, bio-semiotic by Hoffmeyer 1997, eco-semiotic by Kull 1998a, b, and Nöth 1998) is setting the ground for the development of new approaches to overcome the existing limitation.

The need for operational tools to fill the gap between empirical and theoretical sciences operating in the environmental scenario has led to a redefining of the notion of landscape. The landscape is no longer a self-standing physical entity described and interpreted by direct and remote sensing techniques. The landscape is now considered as a mainly human-related entity composed of many processes that are

A. Farina, *Ecology, Cognition and Landscape*, Landscape Series 11,
DOI 10.1007/978-90-481-3138-9_8, © Springer Science+Business Media B.V. 2010

ultimately detected by our senses, thus existing only in our mind (Farina 2000). Such a change from an absolute system of reference to a cognitive referential has represented an important contribution for acquiring new knowledge of the environmental context (Wu and Hobbs 2002). Cognition is a state of knowledge of the surroundings and each organism has this capacity, even organisms without an explicit nervous system (Maturana and Varela 1980, Capra 1996, Hoffmeyer 2008). In this chapter I extend further the role of cognition in order to establish a bridge between evolutionary biology and landscape ecology. By introducing the "eco-field" paradigm I aim to present a conceptual framework for the assessment and interpretation of the biocomplexity of the environment, the "real world."

This vision creates a species-specific space (landscape) in which processes and patterns are perceived differently according to the species. In this way I dismiss the hypothesis of a landscape that is commonly shared and perceived.

In order to advance in this direction it is necessary to integrate ecological principles with the foundations of cognition, semiotic, and autopoiesis theories. Cognition is the capacity of every organism to interact with the surroundings, and for many scientists cognition is equal to life. I agree with this vision. Every living organism has a cognitive capacity, and in vertebrates, especially in man, this capacity is particularly developed and sophisticated.

Cognition can be addressed from different points of view and becomes an interest for ethologists as well as for semioticians, cybernetists, and scientists of complexity.

An extraordinary contribution was made at the beginning of the past century by Jacob von Uexküll (1992 (1934), 1982 (1940)). This scientist elaborated a theory of meaning, from which I will extract the parts most useful in understanding my hypothesis of the eco-field. I intend for "field" to refer to the space of a specific domain. Remember that space, in this context, is a synonym of landscape, and that domain is the necessary space to include all the relationships and integrations for a certain function or process.

Behavior: Perception and Action

Perception forms the basis for every living organism activity. Life is perception or, if you prefer, cognition. In particular, animals have the capacity to enter into relationships with "neutral" objects. A neutral object is an object primarily defined by human perception. It is considered neutral as an assumption that this object can be perceived by all organisms, but this is not true. Von Uexküll easily demonstrated this, reporting an example of a stone that a man can take on the road to chase an angry dog. The stone remains a stone also when it is in the hand of the man, but the meaning changes soon because the stone is used to chase a dog. The stone changes meaning and becomes a carrier of meaning when it enters into contact with a subject. When a neutral object enters into contact with a subject, specific qualities appear, such as a "sitting quality" for a chair, "drinking quality" for a water glass, and "climbing quality" for a ladder.

The quality that an object acquires was called in 1977 by the psychologist James J. Gibson "affordances" (Gibson 1986). He defined affordances as all "action possibilities" latent in the environment, objectively measurable and independent of the individual's ability to recognize them, but always in relation to the actor and therefore dependent on their capabilities (Fig. 8.1).

Every organism enters into contact with objects, and this contact creates a "subjective universe," the Umwelt of von Uexküll. This author distinguished for plants that are organisms without a nervous system a Wohnhülle – a cover of live cells by which they select their stimulus and enter into contact with the "phenomenological world" (Fig. 8.2). A subject that enters into contact with a meaning-carrier can be considered a meaning-utilizer. The Umwelt is considered a closed unit in itself. This vision overlaps largely with the autopoiesis hypothesis of Maturana and Varela (1980), but also the semiotic closure of semioticians. The concordance of these ideas from such different perspectives, ontogenetically independent from each other, reinforces the conviction about the strength of a paradigm based on perception/cognition to be introduced into the realm of ecology. Every organism, according to the resolution of the sense organs, creates a locality or field. For psychologists this cognitive domain is also called cognitive maps but also mental maps or mind maps. They can be defined as "mental processes acting by a series of psychological transformations by which an individual can acquire, code, store, recall, and decode information about the relative locations and attributes of phenomena in their everyday or metaphorical spatial environment." Such maps change character, spatial

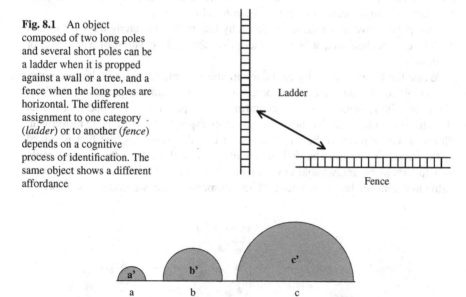

Fig. 8.1 An object composed of two long poles and several short poles can be a ladder when it is propped against a wall or a tree, and a fence when the long poles are horizontal. The different assignment to one category (*ladder*) or to another (*fence*) depends on a cognitive process of identification. The same object shows a different affordance

Fig. 8.2 Every organism has a subjective universe created by the perception toward the external world and powered by the internal autopoiesis. In this case every organism (**a, b, c**) is living in a separate perceptual-cognitive domain or Umwelt (*a', b', c'*)

resolution, and importance in every day life according to human categories (young, adults, male, female) and health condition (Bechtel and Churchman 2002).

Von Uexküll said "everything that falls under the spell of an Umwelt (subjective universe) is altered and newly shaped until it has become a useful meaning-carrier."

It is necessary to distinguish two different components in the meaning-carrier: a "perceptual cue-carrier" and an "effector cue-carrier." In the first case, the perception is the mean attractor (for instance the color of a flower for a butterfly), and the composition of the flower allows or does not allow a butterfly to stop for its nectar. In the second case the effector cue-carrier allows the linking of perception with action.

The theory of meaning offers the possibility of expansion into the concept of the eco-field. Eco-fields can be considered a "spatial configuration meaning-carrier."

The eco-field can be considered the space of existence (domain) of the meaning-carrier when this carrier enters into contact with the subject (Fig. 8.3).

Without spatial configuration, an organism can't recognize the meaning-carrier for a specific function. For instance, a roosting place for starlings is represented by old trees in urban parks. The spatial configuration meaning-carrier is represented by such groups of trees. The inner part of the tree crown is the meaning-carrier for roosting. Both meaning-carriers are necessary to allow a flock of starlings to stop and roost.

For a butterfly like *Argynnis paphia*, the foraging eco-field is represented by a first (spatial configuration) meaning-carrier (context quality): a forest clearing with clumps of *Eupatorium cannabinum*. A second (food quality) meaning-carrier is represented by the flower of this plant. For every function, the coupling of the first meaning-carrier with a second meaning-carrier is necessary. Using the eco-field paradigm, the first meaning-carrier is the eco-field.

This perspective is not far from hierarchy theory, but hierarchy theory does not explain the mechanisms; it is just descriptive about the position of a function or process.

While the Umwelt is easily explained in animals, plants require other premises. In fact plants do not have direct perceptor and/or effector organs; nevertheless, plants are living organisms with cognition and capacities to respond to external stimuli. But in plants there is no ability to select a specific Umwelt using movement. Plants probably react to the environment in two distinct ways: the first is linked to the seeds or spores. The dislocation of successful seeds is the indirect reply to a favorable or unfavorable environment. Success or failure is not a matter of individual choice, but is a matter of the numerousness; we could say lucky or

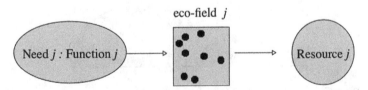

Fig. 8.3 The eco-field can be considered the space configuration meaning-carrier (sensu von Uexküll), indispensable for assigning meaning to every neutral object after a specific function is activated. Needs are connected with resources via active functions and the eco-field interface

unlucky seeds. Survivability and adaptability are in the hands of a collection of seeds. The seed stock is the "individual" that shows sensitivity to the medium. In contrast to animals, plants have a great difference in shape and requirements according to the age and the condition in which they grow. Their morphological plasticity is very high. The shape design is open and not as fixed as for animals. This is probably a compensatory mechanism for the small capacities of plants to move around. Moreover, according to the age of a plant, difference becomes a necessity, and consequently difference is key to the relationship with the environment in terms of nutrients, water, and competitive tolerance. Plants and animals have organs with the capacity to use the meaning-factors that are external to their bodies. Every organism lives in an "operational world" sensu Loeb (1916) that is the representation of sensors (either organs like eyes, tactile sensors, hearing for animals, or chemical reactions by membrane tissues of plant cells). The external world is perceived by organisms in a specific way. An example is given by the spider web. The web of a spider is like a suit for a tailor. But the tailor first measures the body of the customer, and a suit is the hollow shape of the human body. This is not possible for the spider, which nevertheless can build a web to capture flies; thus a web is for a spider the image of the fly! The spider can not measure a fly like a tailor measures a man for a suit, so the web is the idealization of the archetype of fly. This idea opens the way to many interesting arguments about the representation of meaning-carriers, like maps or aerial photographs for a geographer. A vegetation map is an idealized space around us and not the realistic representation of vegetation per se. A map, like a web, captures emergent properties of a space such as heterogeneity and diversity; a spider web captures flies directly. For von Uexküll millions of Umwelts exist according to species and their sensory capacities.

About Ecological Niche and Habitat

The "niche theory" in ecology describes the main functional and structural characteristics recognized by a species (Grinnell 1917, Hutchinson 1957), and represents a foundation of modern ecology. This paradigm has been extraordinarily useful for understanding the evolutionary and adaptive mechanisms by which a species interacts with the environment; nevertheless, this paradigm is unable to account for the mechanisms acting to transfer the information of the life trait into the "real world" and vice versa.

The impressive advancement in information science during the Twentieth Century has emphasized the concept of information itself, which is necessarily associated with its emission, transport, and reception (Shannon and Weaver 1949, Brillouin 2004, Battail 1997). In an evolutionary perspective a large amount of information is carried by resources, which, defined in a broad sense, refers to any environmental variable necessary for the survival of the species, and does not exclusively refer to food and nutrients. Resources can thus include temperature, humidity, and refuges from predators or light for photosynthesis, mental and spiritual elements as well.

The same resource may be used in different ways by different organisms, thus leading to a new definition of "niche" in topological terms according to functional traits that become the niche axes. In such a metric of functional traits, the niche is delimited by the intersection of species environment with the functional circle of resources (von Uexküll 1982 (1940)). Resources can be intercepted at different times and at different scales. This concept emphasizes the dynamics of the resources. Resources are characterized by their temporal availability and their spatial arrangement: generally are not abundant and are heterogeneous in time and space.

In general, the habitat of a species is a physical space under environmental constraints, in which to find food, mating places, refuges, etc. (see Mitchell and Powell 2002 for a discussion and criticism on the habitat definition). There are species that have a broad foraging niche and contemporarily have habitat type restrictions. The classic "habitat" definition based on binary logic (habitat, nonhabitat) can be modified using fuzzy logic to range from a fully suitable to an adverse environment, moving across a broad spectrum of intermediate conditions characterized by different spatio-temporal patterns. These patterns must carry information that is perceived by functional trait sensors. The Umwelt (the "external world" following von Uexküll 1982 (1940)) becomes a subjective representation of the environment according to one organism's mind (Nöth 1998). The different suitability of habitats recall the source-sink model (Pulliam 1988, 1996).

In a way similar to what happens in living organisms, which detect the energy of the physical world through sensory systems that transform sensation into perception via a cognitive process, we assume that functional trait sensors are characterized by specific cognitive properties beyond the "structural pairing" proposed by Maturana (1975). This framework allows us to complete the niche concept by assuming that organisms "perceive" their environment according to a species-specific scale that depends on the functional trait active at a specific time.

The Definition of Eco-Field

When we introduce a new concept associated with a new word in science we aid in the shifting of a paradigm, which leads to an advancement of the science (Kuhn 1962). The habitat concept has also been used for a long time to define the place and the characteristics of the place in which a species can be observed. The habitat concept has also been used outside its original context to describe a type of plant or animal association. In practice, habitat has been recognized as a place with specific characteristics, and the original functional entity has been transformed into a structural entity like a house.

I'll try not to enter into discussion put forward by the opponents of the habitat concept. A huge amount of literature on this subject exists, but it is my intention to prove that the cognitive landscape, such as the summation of all the individual functions related to an eco-field, is a better fit for the investigation of living bio-ecological requirements of a species. This perspective affords extraordinary implications in species management and conservation.

If reality is a matter of organism perception, the perceptual domain can be localized into a space. Space is the phenomenological domain in which life exists, is modified, and moves towards extinction.

We introduce the definition of the eco-field as the ecological space where functional traits, or the niche axes, intercept the resources used to satisfy those needs according to a cognitive perception of the environment (Farina 2000, Farina and Belgrano 2004, 2006, Farina et al. 2005). The eco-field appears as a dynamic interference space within the physical world, in which internal functions and external processes interact continuously as a complex entity, and in which hostile and suitable elements alternate (Fig. 8.4).

The new term eco-field requires further explanation. It is strongly related to the Umwelt concept, but unlike the Umwelt, the eco-field has more ecological implications than the primarily behavioral implications of the Umwelt.

The term "eco" has been introduced to signify that we emphasize the functions, and that these functions are located within a space (sensu Maturana 1975).

The eco-field is a paradigm that is useful for many purposes and is linked with the semiotic concept of sign-signal and "representamen" (see later).

It is curious to find that von Uexküll, Maturana and Varela, and semioticians have described the same phenomenon (the "I" environment versus the "it" environment; see for more details Hoffmeyer 2008) using different approaches that converge toward a unified principle. These researchers have all inspired this synthesis and have stimulated my effort to reshape landscape theory into a more multicomprehensive system.

Life is cognition, the cognition is a semiotic closure, and cognition creates a frame of reference neighboring the Umwelt that is function-specific: the eco-field. New words for new concepts are the basis for a new science.

Following the Natural History paradigm, every ecosystem is characterized by builders (e.g. plants) and users (e.g. animals), but the eco-field allows a much more dynamic perspective that considers all species (plants and animals) to be like agents that play both active and passive roles according to a complex screenplay. In the active role, species enter into competition for resources with other species reacting to the constraints of evolution. In the passive role, species are used as substrate or patch, namely resources for other species-specific eco-fields. For instance, the

a   b

Fig. 8.4 Objects dispersed in a medium become meaning-carriers when a precise space configuration emerges and is recognized. In this case it is evident that the number of items (**a**) is not sufficient to assume per se a meaning but after precise positioning in space the collection of objects shows a specific meaning as a collective property (a smiling face) (**b**)

perception and utilization of a tree in an urban park can be different for squirrel, lizards, birds, or mushrooms. A tree is a patch for these species but alternatively becomes an individual and its Wohnhülle (von Uexküll 1982 (1940), 1992 (1934)) requires light, water, and nutrients.

The Individual Eco-Field and the Definition of the Cognitive Landscape

Originally landscape was considered as a composite structure, a box in which to find processes and organisms. Now that I have defined the eco-field I'll try to incorporate this paradigm into the landscape space (sensu Maturana 1975). Every organism, according to the functions that are active at the time, uses specific sensors (cognitive templates) to detect resources (Fig. 8.5). The resources are detected using a specific eco-field as interface that has a unique dimension and shape according to the specific function.

Finally, if we sum all the eco-fields activated by an individual we have a space differently perceived. Such space is in reality the range of all possible eco-fields; this is the cognitive landscape of that species. It is a landscape that exists around an individual, and it can't exist outside the individual and its characteristics (Fig. 8.6).

Starting from a collection of neutral objects that we could call the "neutral landscape," every individual (plant, animal, but also process) intercepts a part of the

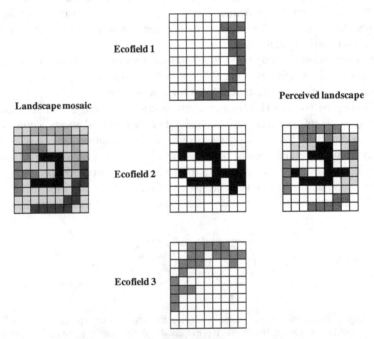

Fig. 8.5 An example of how a species can perceive the complexity of the landscape according to different functions involved in tracking resources in space, time and according to resource quality. This scheme can be applied also to describe how different species use the environmental context

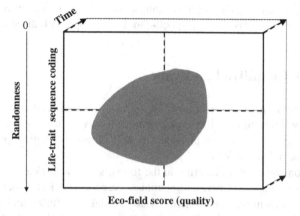

Fig. 8.6 The sequence through which eco-fields are achieved and their quality determine a phase space inside which a species persists, becomes extinct, displaced or adapts. In a phase space all combinations of life-sequence coding and quality are possible. The potential life trait sequences range from full coded (no plasticity and adaptation) to full randomness of sequence

neutral landscape that can only be considered as an implicit entity. An individual decodes from the universe of signals that are produced by a neutral landscape, the ones necessary for its life, and transforms such signals into a specific meaning by using a semiotic process.

The distinction of different meaning-carriers allows species to share the same space at the same time. The same object can assume a different meaning-carrier. A flower can be food source for a butterfly, but also food source for a spider that uses the flower as a pole from which to prey upon pollinator insects (Fig. 8.7).

Fig. 8.7 A flower can assume a different meaning-carrier: for a butterfly a flower is a meaning-carrier of food quality, while for a spider it is a meaning-carrier for predatory position quality. Definitively the affordance of this flower changes according to the agent

The characteristics of the cognitive landscape can not be observed directly by another observer, but only through specific tools that cope with the sensors of focal organisms.

Scoring the Cognitive Landscape

Imagine a specific population of birds living in a forest. Forest is not a homogeneous area by definition, and resources are variously located in the forest. This distribution depends on several physical, biological, and ecological processes that have nonuniform distributions.

Perturbations originating external to the forest, such as hurricanes, logging, and fire can modify shape, history, and spatial aggregation of individual plants. Soil reflects differences in nutrient content, water availability, fungi, and bacteria. All these factors influence the distribution of plants, their individual history, and the collective shape of the forest. Now we investigate the distribution of individuals of a species of bird, for instance the European robin (*Erithacus rubecula*). The distribution of this species in the forest is not homogeneous. There are areas of high and low density. At first sight this pattern is the result of the distribution of resources, and a multitude of studies try to link resource distribution and birds. But if we try to define a resource, we immediately think of food and water. In reality, organisms need more types of "resources" from the environment. For a bird these could be: food, water, opportunity to acquire a mate, nesting site, singing place, resting place, anti-predatory refuge, roosting place, flying space, and feeding space.

It is evident that the meaning of "resource" has been enlarged, and, in addition to biomass availability, we have considered also environmental and social conditions like safety.

This list could be enlarged according to the species considered.

For every need an individual requires a specific "resource" that is intercepted by cognitive mechanisms mediated by specific sensors, such as the eye, ear, nose, tactile sensor, etc. or by "behavior" sensors, such as the choice of roosting place or by cultural and experiential tools.

Every function activated by a specific physiological requirement needs a space in which to find a "structural coupling" with the environment (Fig. 8.8). I have called this space the "eco-field." According to the meaning theory of von Uexküll, the eco-field could be considered "a spatial configuration carrier of meaning" (Fig. 8.4). By definition a space has borders and, consequently, a dimension. Inside the eco-field the conditions to perform that specific function are not isomorphic and heterogeneity creates conditions with a different quality inside a specific eco-field. But assuming that a final unique score can be attributed at every eco-field of a selected species, the quality of the environment in which an individual resides depends on the score of each eco-field. And in turn the score of each eco-field determines the fate of that individual. Let me present another example. If all eco-fields have high scores – except for predation susceptibility which is very high and registers as a low score – survival will be extremely low.

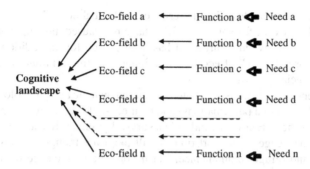

Fig. 8.8 The cognitive landscape is the combined perception and interpretation through the composing eco-fields. This is a unifying principle that can be applied to every species. The activation of a function is necessary to accomplish a need. Every function refers to a cognitive template (cognitive map) that is used for scanning the surroundings to find the corresponding spatial configuration carrier of meaning necessary to intercept the resource

For instance the tree pipit (*Anthus trivialis*) is a bird that, during the breeding season, requires open grassland in which to place the nest, but at the same time it needs a tree from which to jump and to perform a singing flight. If all other eco-fields have a high score but trees are absent, no birds will settle in such a condition. But if trees are present although the food resources are scarce, a pair can establish a territory. Finally, the total scores for all eco-fields will determine the functional condition for a specific individual. A fat individual can be unhappy, and a hungry individual happy. Adaptation mechanisms are active in a real space for precise functions with effects on the genetic selection of individuals in a population. The concept of eco-field demonstrates the possibility of coupling knowledge of the "habitat" with adaptability processes.

A double mechanism is in action for every individual: a priority in selecting a function according to the past history of the actions performed, as well as a differentiated effect according to the qualities of the distinct eco-fields (Table 8.1).

This vision has considerable implications for the management of species and for the reconnection of evolutionary biology to physiological ecology. A species can be resident for a long time in a place, and at a certain moment, can disappear

Table 8.1 The best sequence of food and water depends, for a horse, not only on individual preferences for a certain type of food but on the type of activity carried out before food selection. If the horse was mounted on a hot day, for a long time sequence A could be the best. Case B is for a horse with a moderate activity. Case C is for a horse not in activity. Case D is for a horse after a training set. The sequence can be established according to the memory of the previous actions. A priori action does not exist; the sequence depends on the history

A	B	C	D
Water	Apple	Hay	Fodder
Fodder	Fodder	Fodder	Water
Hay	Hay	Water	Hay
Apple	Water	Apple	Apple

abruptly. This could be described in terms of ecological debt (Tilman et al. 1994). In this case low scores for some eco-fields have not caused any apparent effect on the population, but, when a threshold has been passed for an additional negative factor, the entire "family" of eco-fields no longer has the minimum requirements. As a result, species go extinct.

The sequence of functions (regulated by internal genetic and physiological mechanisms, and by external perturbations) and their eco-field score create a phase-space that delineates the border of the individual survival gradient. If you maintain a wild shrew in a small cage, in a short time it will die, even though you provide fresh food, water, and refuge in abundance. But if you enlarge the cage for instance by providing perforated bricks the shrew can move a lot inside and outside of the brick using the holes and brick tunnels. The shrew can then survive for a long time even if food is scarce. This is an example of different eco-field scores that make a difference. Survival is not the total sum of the eco-field scores, but it depends on the position in which the specified eco-field is ranked. We could define essential and complementary eco-fields, although this distinction can't be absolute only relative and strongly connected with other factors such as seasonality.

Evolutionary Process Epistemology and Cognitive Landscape

The theory of evolution operates assuming that organisms under environmental pressure become extinct or adapt to new conditions. This is quite clear but a rigorous explanation is lacking. Evolutionists state that only the strongest survive, and that mutations create the condition for survival of the best-adapted, or survival of novel genotypes that appear by chance. Assuming a population is heterogeneous in terms of genome and is under a predictable environmental pressure, we can expect the maintenance of such heterogeneity, although we ignore the precise mechanisms. Often in science we state a principle and then we invoke a complicated procedure (statistical or numerical) to demonstrate our assumption. In this way science becomes cryptic and self-explanatory. I believe that life is composed of processes that are basically extremely simple. Complexity appears when we move from individuals to aggregations like populations, communities, meta-communities, systems. The Ptolemaic system, not the Copernican, was complicated. We can't understand a fact if this fact is not distinct from a background.

The cognitive landscape is a novel paradigm, and the principles that result open a new perspective on landscape science.

If a cognitive landscape is the space in which the life web is connected with all the possible relations and interactions of a collection of elements, the organism should have a holistic vision of such a landscape. An organism is not an observer in such a domain, but an active component without the capacity to describe the surroundings as they occur to an observer (man). At this point, it seems difficult to find a good metric able to capture quantitative data from this vision. A possibility is to utilize individual-based "sensory perception" and to move from an operational domain to a descriptive domain. Most metrics used in landscape ecology are related to descriptive domains and not to operational domains.

Later we will discuss this point which is fundamental to a more robust science of landscape in greater detail.

The General Theory of Resources

Every living being is alive for autopoietic mechanisms (Maturana and Varela 1980) that coupled with the (teleonomic) genomic project assures the internal homeostasis and the necessary contact with the external world. Such contact is based on perception and cognition that must be continuously in action to gain the maximum of the information possible. The maintenance of the autopoiesis requires "fuel" that consists of matter (e.g. food), energy (e.g. light), and information (e.g. sound) that we call in a generic way "resources."

Under the paradigmatic framework of the cognitive landscape a special place can be reserved for the General Theory of Resources (Farina, in preparation). This theory states that the deficit of matter, energy, and information that every living system feels, in order to stay alive, namely "needs," represents the major stream of life, the goal of every living being and that it is accomplished through specific functions able to intercept the requested resources, via bio-semiotic mechanisms.

The use of matter, energy, and information is not episodic but requires a continuous process of resource access and resource "re-fueling." For this reason every living being needs a source able to re-source after use. A resource, after every depletion episode must have the capacity to recover and to be available for a successive depletion, entering into a depletion–recovery cycle.

The time lag for resource recovering, when regularly fixed, produces rhythms. In nature rhythms exist as reply to the time interval requested for the regeneration of the resources. The use of resources is connected and timed with the metabolism of organisms, as well.

In order to be localized, every resource must have some regularities that allow a recognition by the users. Regularity means information (see more details in Harms 2006) and when the resource enters into a sign circle with the living being such a relationship is of semethic type (see Hoffmeyer 2008 for more details).

The heterogeneous distribution of resources in space and time and their scarcity require accurate mechanisms of interception to limit the energy demanded to explore the range to intercept resources.

According to the level of indispensability resources can be classified as: compulsory, optional, and marginal. For example, water, food, and air are compulsory. The optional resources include categories of food such as fruit for a diet-generalist species. Marginal are resources like a glass of bier for a man whose availability improves his quality of life a little. Most resources belonging to information are optional or marginal but their shortage produces a bad-being status. The Etruscan shrew (*Suncus etruscus*), the smallest (by mass) known mammal in the world (8 g), requires to stay alive, when caged, contact with the surface of the soil around its body otherwise this micromammal enters a stress status, which in a short time can accelerate towards a death event.

The eco-field described as a spatial configuration carrier of meaning specific for every semethic organism–resource interaction, is a central concept in the General Theory of Resources. The presence of a specific eco-field per se is not enough to guarantee the resources. In fact it is possible to have a suitable eco-field but a lack of the correspondent resource. This condition is well known in ecology and is described as an "ecological trap." For instance, many species of birds nest inside the orchards of vineyards. The orchards probably look very attractive to the birds, but few are successful here because of disturbance by cultivation practices (spraying or green pruning). The "safety" resource is not achieved despite a favorable nesting eco-field, i.e., the dense cover of leaves to hide the nest in and many branches on which to anchor the nesting cup.

By adopting the niche-construction strategy (Odling-Smee et al. 2003) many organisms improve strategies of resource-tracking by adopting a very special type of resource cultivation. For instance, wild boars plough the soil stimulating growth of specific vegetables particularly preferred in their diet.

The abundance of the resources is not a sufficient condition to guarantee well-being, often resources enter into competition with each other and a tradeoff is necessary (Fig. 8.9). For instance the "food" resource and the "safety" resource in many cases enter into competition. Experimental evidences have demonstrated that food resources when located in unsafe locations are ignored producing a malnutrition status in organisms. This vision is otherwise interpreted according to the source-sink model (Pulliam 1988, 1996).

According to the bio-semiotic process involved, resources can be classified into: direct and indirect.

The direct bio-semiotic process involves only one step in signification. The indirect bio-semiotic process requires at least two consecutive steps. Food catching involves only one sign process, but to enjoy the fresh air in an urban park requires one to walk.

Several resources are offered by the same agency. In this case the use of one prevents the use of the second one. For instance biomass (leaves) and refreshment

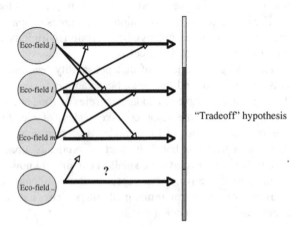

"Tradeoff" hypothesis

Fig. 8.9 It is reasonable to admit that every eco-field is largely interacting with all others and this fact determines a tradeoff in resource tracking

(shade) are two resources that when produced by the same tree can't be used contemporarily. If you use the biomass for feeding the refreshment (resource) disappears. We can say that the two resources enter into competition.

When the same resource is utilized by two or more species, those species enter into competition. This means that each species probably adopts the same eco-field. Evolution acts inside this process reducing the competition for the resources.

When a resource becomes neglected generally the correspondent eco-field vanishes as well. This is the case with most local types of fruits and cultivated seeds. Disuse of the resources produces the disappearance of the correspondent eco-field.

When many resources are neglected at the same time, like the rural products in mountainous areas, all the correspondent eco-fields disappear. If we consider that the summation of all the eco-fields of species creates a cognitive landscape, the transformation of the landscape is the consequence of resource disuse.

When a living being reaches every necessary resource this condition produces a state of well-being. Conversely when some resources are no more available the living being enters into a state of bad-being. Well- and bad-being are important indicators of the life style of our societies.

The general theory of resources can be effectively adopted to interpret human habits but with some precautions. In natural systems resources are tracked inside the home range by every organism in order to reduce the energetic demand. In such a way, an individual species lives where resources exist. This is not always true for humans that have the technological and energetic capacity to gain access to resources far from their home range (Commune, Region, Country, and Continent). Tracking resources from other "territories" requires a lot of energy and this process subtracts such resources from the local populations causing social and economic problems.

Some resources are only potentially and vaguely expected. They can be localized only by adopting cultural tools. This is the case of biodiversity. Biodiversity is a potential source of goods and services for humanity but a lot of cultural information is requested to evaluate this emergent entity.

The General Theory of Resources allows one to incorporate matter, energy, and information into a unique epistemological domain and to create a good context to interpret the ecological landscape. This theory recognizes the resource flux from outside to inside the major teleonomic habits of every organism and it opens a wide window also to the philosophical interpretation of the mechanisms adopted by every living being to stay alive.

Conservation and the Cognitive Landscape

The "cognitive" landscape as a whole may be interpreted as the sum of the different eco-fields of all species. In this case, however, we enter into the domain of description, yet the cognitive landscape is too compacted into a neutral vision to make this possible. Measuring the eco-field for most functional traits is difficult because we usually ignore the range of a single trait, and consequently, lose the metric to be applied. According to the eco-field paradigm, it is nevertheless possible to attempt the separation of the different functional traits using simplified models. This is an important operational procedure used to compare the most important functions of a species and the description of its optimum spatio-temporal distribution which follows those functional traits with the real condition met in a selected area at a given time. It is impossible to know, in advance, which degrees of interaction and priority are associated with the functions selected for the study of a species. However, our approach allows the construction of testable models because we assume that different perspectives may potentially be equally useful for representing the eco-field. The application of the model will provide the results necessary to reconstruct the priorities and the quality of each eco-field in order to predict where a species may have the optima of distribution (see Mitchell and Powell 2002).

This has extraordinary potential for the conservation of species but it is necessary to develop a strict linkage between the genetic variability of a species and the local conditions. The eco-field allows one to investigate habitat suitability for simple functions that we assume are important for the maintenance of a stable population. For instance, the nesting eco-field for a barn swallow (*Hirundo rustica*) is represented by available natural or artificial walls, orientation, height, etc. Let us assume that it is possible to determine the optimal distribution of these functions for swallows. Then it could be possible also to measure the potential nesting-sites in neighboring areas. In a similar way, by counting the composition and abundance of aeroplankton around the nesting places, it could be possible to evaluate the status of a "foraging" eco-field. Following this procedure as a function of time and of the functional traits of the swallow, we can assess the habitat suitability.

The eco-field paradigm could help us to understand how phenotypic plasticity (see also Sultan 2000), is maintained in a population using a combination of values (ranging from completely unsuitable to optimal) of all the eco-fields that an individual of the population has gone through in its life. This procedure has important evolutionary implications because the environmental pressure on the genome of individuals reflects the single eco-field conditions. The genetic variability of a species is determined by different constraints that select the genome better adapted to those conditions. The number of conditions is very large, and the number of possible combinations is even greater. For instance, quality modulation in the "mating" eco-field will influence the density of the future population, but will also influence the foraging eco-field.

Practical issues are raised by conservation strategies aimed at maintaining the integrity of genome potentiality. For instance, in order to maintain the population of *Rupicapra rupicapra ornata* in Abruzzi National Park (Italy), which is an isolated population, would it be better to maintain a high level of habitat quality or would it be better to allow subpopulations to live in suboptimal habitats (estimated on the basis of a demographic model)? What is the effect of the "source-sink" habitat for the maintenance of genetic diversity? It could be of great interest to study the genome variability of subpopulations with different eco-field suitabilities in order to understand what portion of genomic variability is affected by different eco-field qualities.

The eco-field explains very well the complex domain of species life composed of semiotic relationships and contemporarily by genetic adaptation to different conditions. We believe that most genetic variability is produced by external stressors, but this is true only in part. Internal mechanisms that change the rank of importance of the different vital functions operate throughout the eco-field. The phase space is a human representation of a probabilistic range in which the different functional traits operate. Memory, the past history or past sequence of events, is responsible for the present choices. Finally, the eco-field is not a passive representation of reality, but it is an autopoietic process in which an individual is strictly linked with the environment that changes and by which it is modified. We ignore how the different eco-fields rank, but it is probable that some functions are essential and others are additional. If you are very rich, you can perform more things than if you are very poor. The essential functions, such as feeding, drinking, or sleeping, are not very different, but others such as recreation (tourism, concerts, readings) are completely different.

In autopoietic theory, Maturana and Varela only consider the autopoiesis inside an organism, and call all other mechanisms heteropoietic. But see also the point of Zeleny (1996). I consider the recognition of the eco-field essential to autopoiesis and consequently as an integral part of individual autopoiesis. Without the coupling of the eco-field with autopoiesis, this last essential property of the living organisms doesn't exist.

Changes at Individual Resolution: The Application of the Eco-Field Model to Individuals and Systems

According to the eco-field hypothesis every species or process perceives a species/process-specific surrounding (landscape) and any modification of this context produces reactions buffered or enhanced by positive or negative feedback (Farina and Belgrano 2004).

This idea assigns to every specific biological function a physical space in which the resources are intercepted and the external energy is transferred to the internal environment. Scaling properties of each species-specific functional trait create a multitude of eco-fields. The vision of the landscape as a neutral matrix in which every species reacts is superseded by the idea that there exists a species-specific cognitive reality perceived exclusively by that functional trait of the focal species and the totality of the species-specific eco-fields. The Umwelt (sensu von Uexküll 1982 (1940)) becomes the landscape (for that species). This vision allows a formulation of a general theory of the landscape that until now has been split into separate paradigms.

Individual Displacement, Individual Extinction, or Change of Functional Traits: Three Possible Outcomes of Change in the Eco-Field

When a species encounters by chance new environmental conditions three types of responses can occur according to the type of environmental constraint: extinction, geographical displacement, or adaptation.

If the modification occurs quickly and produces an abrupt change in environmental conditions, such as the water table, pH, or salinity, or when an acute pollution event occurs, then individuals may become extinct.

If the modification occurs gradually and the organism has the capacity to compensate or to displace itself (this case is restricted to animals), the individuals can move, searching for a new favorable locality.

Finally, if the change is not dramatic, individuals can "recalculate" the quality of one or more of the eco-fields, and can self-maintain in the same location (Fig. 8.10).

This last possibility is more difficult to measure with the scientific approach due to the relatively new eco-field paradigm (see also von Uexküll 1982 (1940)). However, it is probably the most common and important process for adaptive

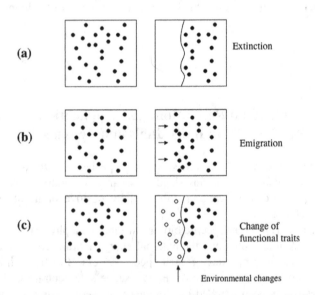

Fig. 8.10 Three possible mechanisms of changes at population resolution of individual reaction to environmental changes. (**a**) A sudden stress moves to extinction a portion of the population. (**b**) Changes are important but the organisms have time (energy) to displace (this case is restricted to mobile organisms). (**c**) The environmental change is not so dramatic and organisms can modify an eco-field over time and maintain at the same time the original position in the living area. A combination of these three possibilities is a realistic mechanism

mechanisms of organisms, and it is a mechanism that greatly contributes to the creation of the ecological debt (sensu Tilman et al. 1994).

This last strategy acts as a redistribution of information among the different eco-fields, exposing species to different levels of environmental constraints. For example, there is the well-known process of phylopatry by which a species persists in an area that has lost most of the favorable characteristics for this particular species. This phenomenon can be explained using the eco-field paradigm in which the fidelity to the natal area has a higher score than the eco-fields related to resources. On the other hand, the phenotypic plasticity is evidence of the capacity of each eco-field to contribute to the stability of individuals and populations in a selected area.

The eco-field paradigm can also be applied to an aggregate system, such as a patch, or a mosaic. It can describe the spatial distribution of an emergent property of the system, such as resilience, stability, resistance, etc.

From this principle it is possible to track backward the effects of environmental changes. In this way, assuming a central place model applied to a forest, the more resilient parts of a forest are the borders, and any change at the border presumably is better buffered than changes in the center of the forest. This is possible if we consider the presence of an active mechanism of adaptation.

A snowstorm damages to a greater extent the interior of a beech forest than the border. This is evident because the border plants have a higher probability to be disturbed by external events, such as wind and snowstorms, than interior plants which are better protected by their numerosity. At the border, plants have trunks that are highly modified by adverse events, and their shape is twisted and, consequently, more resistant to further physical stresses.

Source-Sink Model and Eco-Field

The source-sink model demonstrates how a consistent immigration of individuals from a healthy population may allow a species to perpetuate in a habitat that would otherwise drive the population to extinction (Pulliam 1988, 1996).

Consequently, simple population measures, such as bird abundance during the breeding season, must be associated with a ratio of survival to mortality that is greater than one to be considered robust indicators of habitat suitability.

Outside the breeding season the factors that drive species to select habitats that could have source-sink-like dynamics remain largely unknown.

Some habitats are very attractive for animals but often they are true ecological traps and definitively sink habitats. Animals de facto are deceived by semiotic interfaces that emit a "green light" for a specific habitat. The habitat suitability is decided by an individual on the basis of a specific spatial configuration coupled to a structural configuration. The semiotic process may surprise and force the wrong decision. The tradeoff/integration between the eco-fields remains the major agent of habitat selection, but it's the distribution and abundance of resources that probably drive such source-sink dynamics. The presence or absence of a specific resource like food is discovered only after the "decision" to stay in a place, but after this choice it becomes too late for a species to change "address" and if resources are scarce the risk of local extinction becomes high.

The eco-field hypothesis seems very useful to describe source-sink dynamics because an individual that selects a "sink" habitat does not make a mistake but is deceived by a multiplicity of favorable conditions that remain largely unverified before habitat selection. Each eco-field is acting as an interface between needs and resources, but to couple the needs with resources requires for every organism to experience the structure of that specific eco-field. This means that an individual could stay in a specific place with or without resources. If the second hypothesis is realized the life trait connection with that specific eco-field would be damaged.

Cognitive Landscape and "Neutral Landscape"

It seems like a word game, but the relationships between cognitive landscape and landscape as we intend today, as a matrix or a mosaic, are quite complicated. The observed mosaic is largely the result of the interactions of several cognitive landscapes. Each cognitive landscape has a precise border that could be copied into a larger landscape. Every species enters into semiotic closure with the surroundings, creating new assets and self-changing, but it must be clear that there is no master plan for this. Most of the effects depend on the species considered and on the presence of certain types of species. The recent extirpation of bison in North America has dramatically changed the "neutral landscape" of many other species. At the same time the colonization of Australia by Western societies has changed the mosaic of the arid lands of most of this continent. There is not a master plan acting in nature: Each species operates independently of the others, inside certain limits, and there is not a goal-function to guide the totality of species toward some specific common target. The "human-observed landscape" is a representation of the cognitive landscape, but it could be any of several cognitive landscapes by definition. In conclusion, the landscape is the sum of individual-based cognitive landscapes (IBCL) and species · are the components of a landscape. The landscape is the spatial dimension of perception, and it is the cognitive elaboration of the perception that creates this entity. The environment is considered an entity apart from the species, the space in which species are living, but species are the main producers of the environment. Again we are dealing with a closure, a circle from which it is not possible to escape, but that we have to accept.

Individual-Based Cognitive Landscape and the Societal-Based Landscape: The Human Case

Every human lives in a personal landscape, just like plants, animals, and microbes. But humans have more cognitive capacities, a semiotic niche (sensu Hoffmeyer 2008) dramatically expanded by technological tools. Individual style is a clear manifestation of our capacity to diversify thousands of eco-fields even expanding this concept to the mental processes. Language and culture contribute to expansion and diversification of our semiotic niche.

Visiting an Italian friend's house, we observe a different style of furniture, and, in the kitchen, a different use of food. Opening a conversation on football or on policy, we discover further differences, but when we meet this friend abroad, maybe in Japan, immediately we can find a lot of similarities. We are both "western inhabitants!"

Considering our social aptitude for creating families, groups, societies, and states, we need to understand if the eco-field paradigm can also be used outside the individual. It seems reasonable to believe that we are at the same time individual, clan, group, nation, but for each of these statuses our behavior changes and, consequently, our perception of the surrounding changes as well. Sharing societal characteristics with other persons, we share also a common vision and, consequently, similar or identical eco-fields. In policy, people that take part in a political party have a common vision of the future, and as a result the policy eco-field is similar. We can have different living styles, but when we share the same fanaticism for the same team, we produce the same interference in a specific eco-field.

The human landscapes that we call cultural landscapes are created in a similar way. Considered in a broad sense, people living in a community share a common culture more strongly than we believe. The cultural landscape is an example of societal-based landscape (SBL). It is a landscape created by social mechanisms (by fixed rules) that in turn has been created by a long history of individual interactions shared inside human communities. The cultural landscape differs from the present-time landscape that is produced mainly by economic constraints with little social feedback.

Coming back to the eco-field theory, it is reasonable to imagine that when we feel we are part of an organized entity, our choices are on a cultural basis. We lose our individuality, and we are part of an aggregation. In this way, we perform the processes according to a code processed by an aggregate entity (the society). This procedure is exactly the opposite of the individual-based model. In the latter case, we try to differentiate ourselves to the maximum degree, but in the first case, we try to receive some advantages and to be similar to the other members of the aggregate. In an aggregated entity, our obedience is absolute, and we share common values. We have created many laws to control this type of obedience to the social rules. As a result of this behavior in a SBL we assist in the production of models that are repeated in space and time.

The rules that in the IBCL are the product of sequence and quality (availability of resources), in a SBL are not linked simply to resources and stochastic and genetically influenced processes. Such rules are instead linked to a common vision, like a societal plan that changes at a time scale longer than in IBCL, in which the survivability of the community depends on individual fitness.

The SBL differs from the IBCL by the fact that the SBL is perceived more or less at the same time by groups of persons thanks to the action of a common cultural filter. A SBL has common observed properties that can be understood by the groups without further explanation. Learning rules inside a SBL is a matter of social communication, education, and cultural cohesion.

The IBCL needs to be explained to be understood, and for this reason language is an indispensable tool. Without a language it is not possible to share the IBCL. This is not so important for SBL, at least inside a group. But when we move up and we compare two different SBL, the language that must be used is the language of the policy which is a mix of self-explanation, imposition, and distinctiveness. The SBL produces a geographical entity that we call the "region." The aggregation of different regions produces a country. The region is the most distinct entity that persists for a longer time than a country, at least in the European context. The same political geography of Europe from the Roman Empire's decline to the present time is an extraordinary example of "Inter-regional manipulation."

The Emergent-Properties Based Landscape (EPBL)

When individuals (plants, animals, viruses, or bacteria) aggregate they share common resources and behave differently. A forest, for a flock of birds, assumes a different asset and becomes a system. The behavior of such systems is the object of the science of complexity, and the emergent properties are the basis of the systemic approach. Although it is easy to demonstrate the presence of such properties in a system, it is not clear how and when these properties appear and act. Populations, communities and, more generally, ecosystems have emergent properties, such as stability, resilience, fragility, diversity, etc. that can be empirically evaluated (Muller 1997). The most difficult task is to define such emergent properties that appear only when many entities are in contact with each other (Morowitz 2002).

All these properties are the result of interactions between collections of individuals. If we try to find the geography of such properties, we discover that it is possible to find a gradient in such properties across a space.

Again our goal is to delimit the space in which the emergent characteristics act. While for an individual cognition represents the way to link the internal environment with the external world, the emergent properties are not driven by cognitive processes, but are the result of several inter-individual relationships. Every time we address emergent characteristics, we find it difficult to describe in practice properties like resilience, resistance, and diversity.

Assuming that a population or a community has a spatial distribution, their emergent properties could show such a distribution in terms of a score. The idea that, for instance, resilience is not emerging in a homogeneous way within a community, but that it is heterogeneous, opens the road to novel possibilities in understanding the complexity of the systems.

The central place model assumes an increase in environmental hostility moving from the center to the periphery of the focal entity. At the border of the entity there is a decrease in suitability and an increase in environmental constraints (Fig. 8.11).

If we assume that resilience pertains to a specific domain (resilience domain) and that this domain occupies a space, we could figure out a landscape composed

Fig. 8.11 Representation of
geography of stability in a
hypothetical forest. **a** is the
area with highest stability that
decreases moving from **b** to **c**

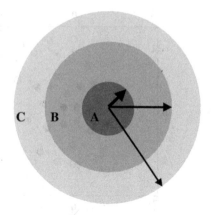

of physical units represented by distinct scores for each emergent property. This idea is very simple but describes the geography of such properties in an efficient way. If properties are not concepts but comprise a functional status of a system, this functional status is not homogeneous because it is not anchored to a homogeneous background.

Moving from individuals to higher order entities, cognition is expressed in terms of areas sensible to energy modification so that different "geographical" parts of a population, community, or landscape play a different role.

For instance, in a flock of birds the individuals that are at the margin are more exposed to predation by hawks, and, considering the probability for all the individuals of a flock to be preyed upon, it is possible to measure this expectation for every individual assuming that every bird maintains its position in the space. If I now express this probability as a risk of predation surface, I can localize individuals that in the flock space are more susceptible to predation. In Fig. 8.12 I have arranged a very simple model of predation based on GIS technology. Individuals that share the center of the flock have a lower risk of predation. This property – it must be clear – is not connected with environmental characters but with the internal organization of the flock. Finally, the eco-field of emergent properties is a special case of IBCL in which properties spring from inter-individual interactions. If, for instance, I eliminate all individuals except one, the predation risk will be completely modified and will no longer be dependent on the position of the individual in the flock because the flock has ceased to exist. When we move from the individual to aggregation of individuals, the interactions are per se elements of organization, and their position makes a difference. Mapping the distribution of emergent properties makes it possible to adopt specific indicators.

Just as for the individual eco-field, all the eco-fields of emergent properties create the Emergent-Properties Based Landscape (EPBL).

This model appears extremely useful in application, considering that to survive, a system must have the highest (absolute) scores for properties like resilience, resistance, novelty, and stability (see Fig. 8.13).

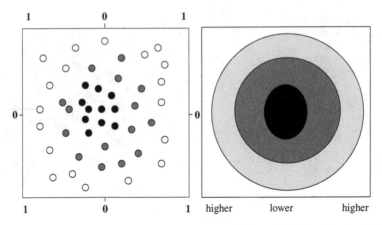

Fig. 8.12 Inside a flock of birds we can distinguish individuals with higher risk of predation (at the border) and safer individuals in the center. This can be expressed as a risk-predation surface

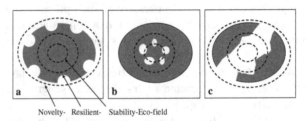

Novelty- Resilient- Stability-Eco-field

Fig. 8.13 According to the eco-field paradigm we can expect different ecofields operating into the entity (e.g. forest), in this case from the exterior to the interior: "novelty", "resilient" and "stability" eco-field. In case **a** the perturbation (clearing) should have secondary effect on the "novelty eco-field" already sensible to changes and environmental constraints. In case **b** the perturbation (clearing) is incorporated by the "resilient eco-field". In case **c** all the three ecofields are involved producing disruptive effects on the described entity

Every property has a geographic distribution according to a gradient, but if we consider (for simplicity) the highest score for each emergent property and the spatial distribution of this property, we could build a mosaic-like surface map that is the emergent properties-based landscape. Some properties can have coincident high scores with other properties, and to localize such areas could be strategic for conserving species and their aggregations.

The differences observed within an entity like a forest or a lichen community are the results of different eco-fields that manipulate energy. The genetic constraint that dominates at the individual level is less evident due to dilution that occurs when a group of such a population, community, or landscape is considered. In such a way we can describe a specific eco-field, such as a "stability eco-field," characterized by an area in which predictability, repetition of patterns, genetic reconnaissance, and coalescence are shaping characters. And in the same way a "novelty eco-field" is active at the periphery, where instability, unpredictability, appearance of new patterns, and genetic diversity are the shaping characters.

The eco-field paradigm is able to link different emergent functions to the geographical space and to provide evidence for the existence of a spatial optimum for every emergent property.

For instance, if a forest is an autopoietic entity, it must have the capacity to create its limits, and thus the ecotones correspond to the limits of such self-organized entities (sensu Jorgensen et al. 1998, Zeleny 1996).

The landscape may be defined not merely as a fixed collection of spatially arranged structures (their total integrated overlap is the "matrix") shared by all species but as a dynamic species-specific and function-related mosaic. This means that each species, according to an active function, may shift within the same landscape, from one eco-field to another and perceive new structures and processes.

The position of a species in the landscape plays a key-role in the evolution of its genetic stock. For instance, at the borders of a landscape (ecotones) the species is exposed to turbulence that may negatively influence some eco-field processes. On the other end, novelties occur at the borders, and novelties are ecologically and biologically important to assure opportunities for the autopoietic mechanisms (Maturana and Varela 1980). The borders appear as places of friction and tension, but they are not the only place of sudden changes in energy levels. In a dynamic similar to the geophysics of tectonic plates, the main constraints occur at the borders of the plates, but fault lines also exist far from the borders where an excess of energy stock appears due to the global set of interactions in the Earth's crust.

The position of a species in the landscape is responsible for the differentiation inside a population that may produce both genetic differentiation into subpopulations inside a community, as well as different coalescence level. The landscape represents the ensemble of all eco-fields of each species, which is the result of bio-semiotic processes. The perception of the landscape is de facto the merging perception of the component eco-fields.

Eco-Semiotic and Cognitive Landscape

Shape, size, and spatial arrangement of patches composing a geographical landscape have an effect on the ecology of organisms, populations, and communities, as extensively described (Klink et al. 2002, Wu and Hobbs 2002, Farina 2006 for a review), but the mechanisms by which organisms use such patterns to identify specific resources like food, roosting, or mating sites remain largely obscure.

For instance, spatial and structural attributes of a landscape like perforation, dissection, fragmentation (sensu stricto), shrinkage, and attrition produced by logging, ranching, or urban development inside forest areas (Bogaert et al. 2004, Levey et al. 2005) affect distribution, abundance, and behavior of organisms, and it becomes relevant to investigate the mechanisms that link such patterns with the habits of organisms (Farina 2008).

In order to fill this epistemological gap, we have introduced the hypothesis that landscape is not only a geographical and ecological space but an agent that functions like a semiotic interface between resources and organism's needs. And according to this perspective, most of the principles on which the landscape ecology is based (Risser et al. 1984, Turner et al. 2001) become insufficient for understanding the ecology of organisms in space.

Contemporarily, cognitive ecology has produced a lot of theoretical disputes, field observations, and experimental evidence (Garling and Evans 1991, Real 1993, Bennett 1996, Allen and Bekoff 1997, Dukas 1998, Nuallain 2000, Chittka and Thompson 2001, Shettleworth 2001, Grubb 2003), but, again, for a better understanding of the mechanisms by which cognitive processes intercept the recognized objects, it is necessary to use the fundaments of sign theory (Peirce 1955, Eco 1975) and their extension to the biological domain (Kull 1998a, Nöth 1998, Barbieri 2003, Hoffmeyer 2005, Nöth 2005, Favareau 2006, Barbieri 2008, Favareu 2008, Hoffmeyer 2008, Kull et al. 2008).

Recently Kull et al. (2008) discussed the necessity to introduce a bio-semiotic approach to the study of the living systems. In particular they individualized important compartments of biology in which the bio-semiotic approach is important like animal communication, bio-semiotic processes in ecosystems (plant–pollinator interaction), the immune system, and neurosemiotics, etc.

In particular, the eco-semiotic framework, defined by Nöth (1998) as the semiotics paradigm transferred into the ecological domain, could help us to understand the mechanisms by which species connect their autopoietic characteristics to the environmental context of the geographical landscape.

Every organism receives signals from the external nature (Hoffmeyer 1996) and at the same time maintains a homeostatic internal "quasi" equilibrium by autopoietic processes (sensu, Maturana and Varela 1980) that activate a certain level of internal insulation.

Probably, two different mechanisms have to be invoked to explain the ecology of the species: an involuntary environmental constraint and a selected voluntary cognitive action.

In the first case according to the individually based perceptional landscape (Farina et al. 2005) the spatial arrangement of the landscape mosaic, like woodland, fields and urban areas, affect organisms without the mediation of explicit cognitive mechanisms. This process is the result of all information (sensu, Stonier 1990, 1996) that an organism intercepts from the surrounding world without a cognitive explicit elaboration. The environmental constraints (e.g. temperature, humidity, air turbulence, noise, water deficit, food availability) operate directly on the physiology or indirectly by exposing organisms to a modified probability of "fitness performance" (e.g. increased predation or starvation risk).

In the second case, organisms perceive physical gradients from their surroundings and convert them into signs by a semiotic (cognitive) process that incorporates meaningful information (sensu, Menant 2003) inside its inner domain, while the surrounding context is transformed into its "private universe" or Umwelt (sensu, von Uexküll 1982 (1940)).

Often the interactions between organisms are so specific that they can be considered semethic interactions (from the Greek semeion = sign + ethos = habit) defined by Hoffmeyer (2008) as: "Whenever a regular behavior or habit of an individual or species is interpreted as a sign by some other individuals (conspecific or alter-specific) and is reacted upon through the release of yet other regular behaviors or habits, we have a case of semethic interactions."

With the increase of their cognitive mechanisms, organisms transform themselves from "objects into subjects of intentionality," and a growing asymmetry in information exchange appears. Specific details of the surrounding world become consistent with the IBCL hypothesis (Farina et al. 2005) that represents the elaboration of the perceived elements through experience, learning, and culture.

Signs from the Landscape

The bio-physical world is full of cues (visual, acoustic, olfactory, electrical, magnetic), that are intercepted as signals for specific goals by organisms. To convert such signals into signs (meaning), a cognitive mechanism is necessarily hypothesized (Hoffmeyer 2005, 2008) (Fig. 8.14).

A signification model based on a triadic Peircean relationship between an object, a sign vehicle (representamen) and an interpretant (Peirce 1955) seems a good candidate (Fig. 8.15). A sign is the product of the association between an object, a sign vehicle, and an interpretant. The interpretant is produced in the mind of the organism and this process is not only limited to humans but can also be extended to other organisms. For a review on the origin and evolution of sign theory see Eco (1975), Nozawa (2000), Sharov (2002), Vehkavaara (2005) and Favareau (2006).

Sign processes reduce the uncertainty to which an organism could be exposed to from the external world and it seems to be an evolutionary short-cut common to the animal and plant realms. When the sign theory is extended to the landscape domain, the individual-based cognitive landscape could be considered a semiotic interface between resources, where organisms function as interpreters.

Every organism, in order to have access to "resources," must interact with the geographical landscape that provides species-specific signals from the spatial configurations of specific objects. Without such (eco-)semiosis, the maintenance of life is impossible, as argued by Hoffmeyer (2005) (Fig. 8.16).

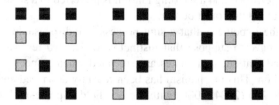

Fig. 8.14 From a geographical (neutral) landscape bio-semiotic processes can extract spatial configurations that have a meaning for a specific function. In this example the geographical landscape is composed of 12 elements that can be aggregated differently producing a I or a T or a 0 accordingly

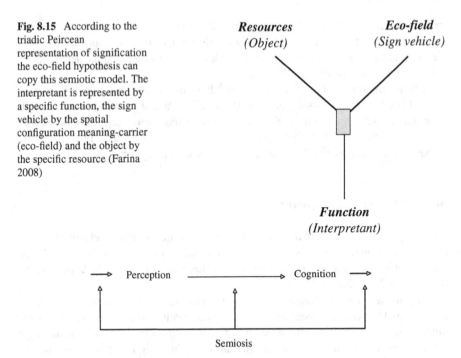

Fig. 8.15 According to the triadic Peircean representation of signification the eco-field hypothesis can copy this semiotic model. The interpretant is represented by a specific function, the sign vehicle by the spatial configuration meaning-carrier (eco-field) and the object by the specific resource (Farina 2008)

Fig. 8.16 The relationship between perception, semiosis, and cognition. The semiotic process is in action across all the interactions of an individual or species with the external world

The integration of the sign theory with the landscape definitively requires that the signals are not linked to a specific object, like a tree or an animal, or to a behavioral process, but that they are the result of the interpretation of different spatial configurations of objects.

Such spatial configurations may be represented by the spatial distribution of trees in a forest, the nectar-dispensing flowers in a meadow, or the extension (e.g. the size of a forest clearing) and shape (e.g. the fractal attribute of a forest margin) of a suitable habitat, but many others could be described. For instance, the woodlark (*Lullula arborea*) requires a minimum number of clearings inside a forested area to select a nesting territory, and a winding margin is preferred to a straight one by deer during grazing along the forest border.

To support the hypothesis that a sign process is indispensable in animal communication we have to suppose that instinctive or learned templates exist in the animal mind and that such templates are activated when a specific internal function is switched on. This mechanism has been recently described and explained by Farina and Belgrano (2004, 2006) with the eco-field hypothesis: For every (vital) function (foraging, roosting, mating, territorial behavior) that requires an external context (surroundings) to be performed, a spatial arrangement of objects (trees, shrubs, other organisms, predators, food) is actively searched. This configuration functions like an interface to localize the specific resource.

The activated function can be processed after such an arrangement has been found (see Weiss and Papaj 2003). Signals from the landscape are transformed into signs by cognitive processes only when a specific function is active, otherwise such signals are not carriers of meaning and organisms are surrounded by a neutral landscape (sensu, Farina et al. 2005).

Following this reasoning, the eco-field hypothesis can be incorporated into the framework of the sign theory in which a triadic relationship links together the resources (object) with the function (interpretant) by the eco-field (sign vehicle).

The "Dimension" of the Signs

It is reasonable to hypothesize that the degradation of a signal, like a sound changing in intensity across distance, may be transformed into different types of signs, accordingly. For instance, acoustic cues like a song strophe or an alarm call may be effective when organisms are close to the source. However, in this signal degradation, an organism must apply a metric in order to evaluate the threshold of the sign efficacy. We could call such a type of sign a "dimension-dependent sign" that can be distinguished from a "nondimension dependent sign," like symbols or icons. Dimension-dependent signs may be acoustic, visual, or functional, like the amount of food in a place. Such signs are under the control of a process that utilizes a cognitive template for an immediate comparison, like the search image mechanism. For instance, when an organism is hungry, a sign of food presence is quickly identified, even when a low level of food availability (the signal) is offered.

There is evidence of sophisticated mechanisms in sign processing in terms of pattern recognition and learning capacity. For instance, the exposition to different live predators produces, in Black-capped Chickadees (*Poecile atricapilla*), different alarm calls (Templeton et al. 2005) and this demonstrates the capacity in this species to discriminate signals.

In many insects foraging for nectar or pollen, a learned fidelity to a plant species has been interpreted as a strategy for optimizing energy in food searching. For instance, Goulson et al. (1997) observed 85% flower fidelity in the butterfly *Thymelicus flavus*, which ignored the presence of other sources of food during the searching behavior. This process is a clear indication of the capacity not only to learn but also to modify the meaning of the sign.

The Soundscape: A Peculiar Landscape

The term soundscape has been used by Ray Murray Schafer (1977) to indicate the spatial and temporal distribution of sound in natural and human-modified ecosystems.

The soundscape is a highly dynamic field of energy and information that has a quite low persistence in time and space. Every acoustic activity from natural processes like a thunderstorm, a water fall, a marine wave or wind and from organisms (bird song, mammal vocalizations) create a spatial acoustic map that can be used by organisms to explore the surroundings.

A huge amount of literature on song, alarm and vocal communication describes functions and patterns of this universal semiotic mechanism (Hopp et al. 1998). But the soundscape can also be considered a peculiar "scape," an organized energetic and informative field used by organisms to increase the semethic connection and to expand the semiotic niche (sensu Hoffmeyer 2008), and to partition communication space between competitive species (Luther 2008).

Hoffmeyer recently wrote: "The idea behind the concept of the semiotic niche was to construct a term that would embrace the totality of signs or cues in the surroundings of an organism – signs that it must be able to meaningfully interpret to ensure its survival and welfare." The semiotic niche includes all the components of the ecological niche but in addition allows organisms to distinguish relevant from irrelevant items and threats and represents the "externalistic counterpart" to the Umwelt concept (Hoffmeyer 2008).

Sounds are the result of a huge energy investment as demonstrated by empirical and experimental evidences (Truax 2001). Communication is seen as the main reason for acoustic cues. But recently we have posed further questions on this subject. Despite a great amount of literature on this subject, especially for birds, marine mammals, bats, and insects (Hopp et al. 1998) largely unknown remain the use of sounds for other purposes. The interesting hypothesis that sounds are used not only to communicate but also as indicators of resources or to locate possible competitors and predators seems confirmed by our recent unpublished results. Using a matrix of sound recorders it has been possible to collect information about intensity and frequency of bird acoustic activity and by a complex elaboration to extract the distribution of sounds across a spatial and temporal surface. We call such a surface a soundscape in which it is possible to observe the tridimensional shape of selected frequencies. With this methodology it is possible to investigate the distribution in space and time of sounds and to interpret the complex patterns created by interspecific interactions. The use by an individual of the spatial distribution of sound perceived around that individual could represent an acoustic eco-field when we assume that some resources can be localized using such a strategy. For instance, the alarm call of a tit, informs other individuals or other species about the presence of a potential risk (man or predator). At the same time, the call of a chaffinch could be used by another chaffinch to find a foraging place. Bird song is used to delimit a breeding territory but also to attract females and chase intruders.

Conserving Natural Soundscapes

The natural soundscape is the result of vocal activity of many organisms (vertebrate and invertebrate, terrestrial and aquatic) that use this mechanism to communicate and to adapt themselves to an unpredictable world. The soundscape is the result of the overlap of biophonies, geophonies, and anthrophonies (Fig. 8.17).

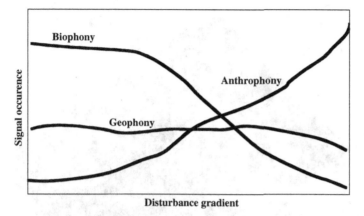

Fig. 8.17 The potential trend of biophony, geophony, and anthrophony along a gradient of disturbance (after Napoletano 2008)

The conservation of the natural soundscape allows one to conserve and to preserve habitat and biodiversity. Bernie Krause, an American bioacoustician that coined the term, "biophony" (Krause 1987), warns about the risk that sounds from many habitats "are forever silenced" (Krause 2002) and that the increasing human clamor produces dangerous consequences on community equilibria. He reported evidences that the frog chorus when interrupted by strong sounds like the passage of an aircraft, allow predators to intercept easily the singing frogs. We have a suspicion that the poverty of beech forest along the Apennines (Italy) is related to the continuous noise of airplanes crossing the region (Fig. 8.18). The symphony of Nature is the result of the vocalization of creatures in relationship to one another and this symphony has several consequences on the vital habits of many species. Krause wrote: "This biophony, or creature choir, serves as a vital gauge of a habitat's health. But it also conveys data about its age, its level of stress, and can provide us with an abundance of other valuable new information."

Soundscapes have to be considered not only as an important aesthetic component of our living system but also as a true resource (spiritual and symbolic for humans, and as adaptative mechanisms for many other species). In our genotype probably persist the ancestral mechanisms that allow us to distinguish signals of dangerousness, signals of a safe place, and signals of changes of a specific situation. The sudden silence that follows when a predator appears in a natural scenario is immediately considered a sign of danger also by inexperienced people. On the contrary the rhythmic biophony of a quiet forest immediately evokes a feeling of safety.

The soundscape is not a separate component but a vital part of our fragile biological surroundings. Human intrusion into natural systems and the use of fossil fuels for energy have reduced biophony and geophony and advanced anthrophony. This trend can be observed moving from natural systems toward urbanized systems. Animal communication remains strongly affected by the increase of anthrophony that can

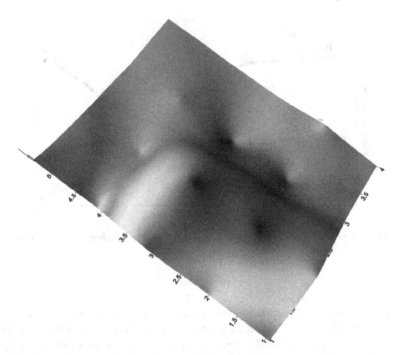

Fig. 8.18 A soundscape of a bird community living in a beech forest of the Northern Apennines, during the breeding season. Peaks represent the area with a maximum of acoustic activity recorded by using 20 digital recorders 100 m apart from each other

reduce such traits with dramatic influences on population survival and community coalescence.

The soundscape represents a new and challenging argument shared by ecological, bioacoustic and eco-semiotic approaches. The soundscape is an energy dimension in space, time, and energy well presented by FFT (Fourier Fast Transforms) that enable one to move from a temporal domain of sound to the frequency domain and allow sound to be "visible." Spectrograms are the representation of frequencies in time and intensity. A fourth dimension is represented by the location of each sound source. Following the Farina methods (Farina and Morri 2008), Morri (2008) it is possible to quantify sound in terms of frequency (Hz)xy and intensity (dB)xy at the different locations (xy) and to interpolate the data on a geographical map by using interpolation procedures from geostatistic and GIS algorithms (Fig. 8.18).

The distribution of sounds across a landscape (the soundscape) represents a source of information that can be used by organisms to localize resources, to avoid predators, to reduce competition.

The soundscape contributes to the functioning of several eco-fields such as the mating eco-field, foraging eco-field, safety eco-field, etc. The acoustic cues and their spatiality probably guide information in a very efficient and fast way. The short memory of the physical acoustic phenomena obliges organisms to repeat messages

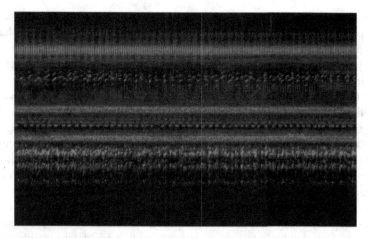

Fig. 8.19 Spectrogram from a Congo forest clearing soundscape. The acoustic (semiotic) niche is shared by at least by six types of organisms. Sound record by acoustic globe at 18.15 in a forest clearing (Bai Hokou, WWF research station in a primeval forest in the Dzanga Sangha Reserve – Central African Republic. 2°51′10.97″ N 16°27′52.43″ E, courtesy of David Monacchi, University of Macerata, Italy)

several times a day according to the presence of other overlapping and degrading sounds (Fig. 8.19).

The well-known Lombard effect produces an increase of loudness in many song birds living in landscapes dominated by anthrophonies.

The regularities of insect, amphibian, and bird sounds are very informative processes that act as key-stone elements in a sea of silence.

Cognitive Landscape Versus Geographical Landscape

According to the approach that we intend to utilize, the landscape appears at the same time as a geographical entity, an aesthetic surface (Barrett et al. 2009), or a mental representation (Gould and White 1974, Lynch 1976, Gibson 1979, Kaplan and Kaplan 1989, Bourassa 1991, Appleton 1996, Ingold 2000).

Probably every approach in some way makes sense, and the difficulty consists of integrating each perspective into a unitarian paradigm.

The hypothesis that landscape is a semiotic interface between organisms and resources (as a result of the integration of all the species-specific eco-fields) is not in contrast with the other ecological theories but simply introduces a new way to evaluate characters that increase our confidence and knowledge about this complex subject (Farina 2008).

Resources like food, water, refuges, nesting or roosting places, social opportunities, amenity and security, sense of place, social identity are not uniformly distributed and accessible; their scarcity and elusiveness in time and space requires some searching efforts to be located. Searching for resources is not a behavior randomly performed by organisms but it is guided by cognitive processes that use mental maps (search images) and trained behavior in order to optimize the efforts made by a memorization process (Sutherland and Gass 1995, Etzenhouser et al. 1998, Griffiths and Clayton 2001, Chittka and Thompson 2001).

Sign theory integrated into the eco-field hypothesis contributes to increasing the comprehension of the mechanisms by which species interact with the surroundings, tracking resources and adapting to the specific ecological niche (see Odling-Smee et al. 2003). This theoretical body explains the mechanisms by which a physiological (e.g. hunger, thirst) or a psychological (e.g. safety, happiness, spirituality) necessity is satisfied through the transformation of a perceived signal into a sign vehicle, and finally into a specific meaning.

In a continuous switch (on/off and vice versa) of functions, that are related to the physiological bodily constraint, we expect a continuum of sign processes that fires like an extended brain from the surroundings that could depend either on the internal status of organisms or on the availability of resources located in unknown surroundings.

The integration of the sign theory into the eco-field hypothesis, using vital functions as "trait d'union" (the interpretant), allows a new representation of the landscape that becomes a cognitive entity composed of abiotic and biotic components afforded (sensu, Gibson 1979) differently according to the interacting species and the related functions (Fig. 8.20).

Finally, a common theoretical framework around the landscape as a functional entity, a biotext (sensu, Kull 2002), would have a positive impact on the development of an interdisciplinary coalescence in the field of conservation, resource management, and planning. The theory of the cognitive landscape as a semiotic interface between resources and organisms can be accepted by geographers, ecologists, landscape architects, anthropologists and environmental psychologists, planners

Fig. 8.20 This is an experimental foraging site for the European robin (*Erithacus rubecula*). Food (millet, sunflower, and mealworms) is provided by the human observer and the white panel plus the container are considered a neutral context in which food is served. Moreover the stopwatch placed at the center of the foraging place has no apparent influence on the foraging behavior. The robin observed by a video camera (Wing Scape Bird Cam), does not show embarrassment (**a**) and it uses the stopwatch glass (**b**) as a platform to capture the offered food

and decision makers. Simulations based on the performance of distinct functions necessary for humans and nonhuman organisms to have access to resources and consequently to establish, expand, or maintain populations are today possible using this theoretical body. Specifically, overlap, conflict, or synergic effects of environmental policies and applications could be tested accordingly, thus avoiding useless discussions about nature, the spatial and temporal delimitation of the landscape, or the inclusion or the exclusion of mental processes in the landscape domain.

Discussion

In this presentation we have defined a new concept, the eco-field, as an operational approach to integrate biocomplexity within the frameworks of landscape and evolutionary biology. The eco-field is important for short term as well as long-term evolutionary processes. At a short-term scale it affects the survival of individuals, and over the long term it drives the arrangement of the genome quality of a species. The eco-field paradigm accounts in a good way for the "source-sink" model (Pulliam 1988, 1996), and incorporates the concept of fuzziness in the distribution of organisms.

We observe the landscape created by our eco-fields but we are urged to search for appropriate techniques to investigate how the landscape is perceived by other organisms to assure their own survival. The human eco-fields are more numerous than those of any other organism because beyond the life traits fundamental for the survival of *Homo sapiens* we have other eco-fields as a product of our mental activity, history, and culture (Naveh 1995). Much of our landscape vision is strongly affected by culture in terms of aesthetic appreciation, although other eco-fields are equivalent for the fitness of our biological components.

The interaction between all eco-fields of all organisms can be finally represented with the ecosystem vision. This vision is becoming increasingly important with the advancement of the concept of biocomplexity (see Lewin 1999). It opens up new ways to investigate the mechanisms that produce emergent properties of ecological systems as the result of nonlinear, causally sorted interactions of different eco-fields. The eco-field paradigm is a representation of a part of complexity. It is not a mechanism itself, but a scheme that emphasizes interactions and information processing in the ecosystem.

Suggested Reading

Bissonette, J.A. and Storch, I. (eds.) 2007. Temporal dimensions of landscape ecology. Springer, New York.
Bonabeau, E., Dorigo, M., and Theraulaz, G. 1999. Swarm intelligence. From natural to artificial systems. Santa Fe Institute, New Mexico.

Garling, T. and Evans, G.W. (eds.) 1991. Environment cognition and action. An integrated approach. Oxford University Press, New York.

Chittka, L. and Thompson, J.D. 2001. Cognitive ecology of pollination. Cambridge University Press, Cambridge.

Dukas, R. (ed.) 1998. Cognitive ecology. The evolutionary ecology of information processing and decision making. The University of Chicago Press, Chicago, pp. 405–409.

Grimm, W. and Railsback, S.F. 2005. Individual-based modeling and ecology. Princeton University Press, Princeton, NJ.

Healy, S. (ed.) 1998. Spatial representation in animals. Oxford University Press, Oxford.

Hopp, S.L., Owren, M.J., and Evans, C.S. (eds.) 1998. Animal acoustic communication, Springer, Berlin.

Sebeok, T.A. 1968. Animal communication. Indiana University Press, Bloomington, IN.

Sebeok, T.A. and Umiker-Sebeok, J. (eds.) 1992. Biosemiotics. The semiotic web 1991. Mouton de Gruyter, Berlin.

Stephens, D.W. and Krebs, J.R. 1986. Foraging theory. Princeton University Press, Princeton, NJ.

Swingland, I.R. and Greenwood, P.J. 1984. The ecology of animal movements. Clarendon Press, Oxford.

Truax, B. 2001. Acoustic communication. Ablex Publishing, Westport, CT.

References

Allen, C., and Bekoff, M. 1997. Species of mind. MIT, Cambridge, MA.

Appleton, J. 1996. The experience of landscape. Wiley, New York.

Barbieri, M. 2003. The organic codes. An introduction to semantic biology. Cambridge Academic Press, Cambridge.

Barbieri, M. 2008. What is biosemiotics? Biosemiotics 1: 1–3.

Barrett, T.L., Farina, A., Barrett, G.W. 2009. Positioning aesthetic landscape as economy. Landscape Ecology 10.1007/s10980-009-9326-z.

Battail, G. 1997. Theorie de l'information. Masson, Paris.

Bechtel, R.B. and Churchman, A. 2002. Handbook of environmental psychology. Wiley & Sons, New York.

Bennett, A.T.D. 1996. Do animals have cognitive maps? The Journal of Experimental Biology 199: 219–224.

Bogaert, J., Ceulemans, R., and Salvador-Van Eyesenrode, D. 2004. A decision tree algorithm for detection of spatial processes in landscape transformation. Environmental Management 33: 62–73.

Bourassa, S.C. 1991. The aesthetics of landscape. Belhaven Press, London.

Bradbury, R.H., Green, D.G., Snoad, N. 2000. Are ecosystems complex systems? In: Bossomaier, T.R. and Green, D.G. (eds.), Complex systems. Cambridge University Press, Cambridge, pp. 339–365.

Brillouin, L. 2004. Science and information theory. Dover Publications, Mineola.

Capra, F. 1996. The web of life. Doubleday-Anchor Books, New York.

Chittka, L. and Thompson, J.D. 2001. Cognitive ecology of pollination. Cambridge University Press, Cambridge.

Cilliers, P. 1998. Complexity & postmodernism. Understanding complex systems. Routledge, London.

Dukas, R. 1998. Cognitive ecology: Prospects. In: Dukas, R. (ed.), Cognitive ecology. The evolutionary ecology of information processing and decision making. The University of Chicago Press, Chicago, pp. 405–409.

Eco, U. 1975. Trattato di semiotica generale. Bompiani, Milano.

Etzenhouser, M.J., Owens, M.K., Spalinger, D.E., and Murden, S.B. 1998. Foraging behavior of browsing ruminants in a heterogeneous landscape. Landscape Ecology 13: 55–64.

Farina, A. 2000. Landscape ecology in action. Kluwer, Dordrecht, NL.

Farina, A. 2006. Principles and methods in landscape ecology—Toward a science of landscape. Springer, Dordrecht.

Farina, A. 2008. The landscape as a semiotic interface between organisms and resources. Biosemiotics 1:75–83.

Farina, A., and Belgrano, A. 2004. The eco-field: A new paradigm for landscape ecology. Ecological Research 19: 107–110.

Farina, A., and Belgrano, A. 2006. The eco-field hypothesis: Toward a cognitive landscape. Landscape Ecology 21: 5–17.

Farina, A., Bogaert, J., and Schipani, I. 2005. Cognitive landscape and information: New perspectives to investigate the ecological complexity. BioSystems 79: 235–240.

Farina, A. and Morri, D. 2008. Source-sink & eco-field: Ipotesi ed evidenze sperimentali. Proceedings SIEP-IALE, Bari.

Favareau, D. 2006. The evolutionary history of biosemiotics. In: Barbieri, M. (ed.), Introduction to biosemiotics. Springer, Berlin, pp. 1–67.

Favareau, D. 2008. The biosemiotic turn. Biosemiotics 1:5–23.

Garling, T. and Evans, G.W. (eds.) 1991. Environment cognition and action. Oxford University Press, New York.

Gibson, J.J. 1979. The ecological approach to visual perception. Houghton Mifflin, Boston.

Gibson, J.J. 1986. The ecological approach to the visual perception. Erlbaum, London.

Gould, P., and White, R. 1974. Mental maps. Allen & Unwin, London.

Goulson, D., Ollerton, J., and Sluman, C. 1997. Foraging strategies in the small skipper butterfly, *Thymelicus flavus*: When to switch? Animal Behaviour 53: 1009–1016.

Graham, M.H. and Dayton, P.K. 2002. On the evolution of ecological ideas: Paradigms and scientific progress. Ecology 83(6): 1481–1489.

Griffiths, D.P., and Clayton, N.S. 2001. Testing episodic memory in animals: A new approach. Physiology & Behavior 73: 755–762.

Grinnell, J. 1917. The niche-relationships of the California thrasher. The Auk 34: 427–433.

Grubb, T.C. jr. 2003. The mind of the trout: a cognitive ecology for biologists and anglers. The Wisconsin University Press, Madison, WI.

Haber, W. 2002. Ethics and morality in the sciences. INTECOL, Seoul.

Harms, W.F. 2006. What is information? Three concepts. Biologicasl Theory 1(3): 230–242.

Hoffmeyer, J. 1996. Signs of meaning in the universe. Indiana University Press, Bloomington, IN.

Hoffmeyer, J. 1997. Biosemiotics: Toward a new synthesis in Biology. European Journal for Semiotic Studies 9: 355–376.

Hoffmeyer, J. 2005. Biosemiotik. En fhandling om livets tegn og teggnenes liv. RIES.

Hoffmeyer, J. 2008. The semiotic niche. Journal of Mediterranean Ecology 9:5–30.

Hopp, S.L., Owren, M.J., and Evans, C.S. (eds.) 1998. Animal acoustic communication. Springer, Berlin.

Hutchinson, G.E. 1957. Concluding remarks. Cold Spring Harbor Symposium on Quantitative Biology 22: 415–427.

Ingold, T. 2000. The perception of the environment. Routledge, London.

Jorgensen, S.E., Mejer, H., and Nielsen, S.N. 1998. Ecosystem as self-organizing critical systems. Ecological Modelling 111: 261–268.

Kaplan, R., and Kaplan, S. 1989. The experience of nature. Cambridge University Press, Cambridge.

Kauffman, S. 1993. The origins of order. Oxford University Press, New York.

Khun, T.S. 1962. The structure of scientific revolution. University of Chicago Press, Chicago, IL.

Klink, H.-J., Potschin, M., Tress, B., Tress, G., Volk, M., and Steinhardt, U. 2002. Landscape and landscape ecology. In: Bastian, O. and Steinhardt, U. (eds.), Development and perspective of landscape ecology. Kluwer, Dordrecht, pp. 1–47.

Krause, B. 1987. The niche hypothesis: How animals taught us to dance and sing. Whole Earth Review.

Krause, B. 2002. The loss of natural soundscapes. Earth Island Journal 27-29.

Kull, K. 1998a. Semiotic ecology: Different natures in the semiosphere. Sign Systems Studies 26: 344–371.

Kull, K. 1998b. On semiosis, Umwelt, and semiosphere. Semiotica 120(3/4): 299–310.

Kull, K. 2002. A sign is not alive—A text is. Sign System Studies 30: 327–335.

Kull, K., Emmeche, C., and Favareau, D. 2008. Biosemiotic questions. Biosemiotics 1: 41–55.

Levey, D.J., Bolker, B.M., Tewksbury, J.J., Sargent, S., and Haddad, N.M. 2005. Effects of landscape corridors on seed dispersal by birds. Science 309: 146–148.

Lewin, R. 1999. Complexity. Life at the edge of chaos. The University of Chicago Press, Chicago.

Loeb, J. 1916. The organism as a whole: from a physicochemical viewpoint. Knickerbocker Press, New York.

Luther, D.A. 2008. The evolution of communication in a complex acoustic environment. Unpublished PhD Thesis, Chapel Hill.

Lynch, K. 1976. Managing the sense of a region. MIT, Cambridge, MA.

Manson, S.M. 2001. Simplifying complexity: A review of complexity theory. Geoforum 32: 1405–1414.

Maturana, H.R. 1975. The organisation of the living. A theory of the living organisation. International Journal of Man–Machine Studies 7: 313–332.

Maturana, H.R. and Varela, J.F. 1980. Autopoiesis and cognition. The realization of the living. Rediel Publishing Company, Dordrecht, Holland.

May, R. 1974. Biological populations with non-overlapping populations: Stable points, stable cycles, and chaos. Science 186: 645–647.

May, R. 1976. Simple mathematical models with very complicated dynamics. Nature 261: 459–467.

May, R. 1986. When two and two does not make four: Non-linear phenomena in ecology. Proceedings of the Royal Society B228: 241.

Menant, C. 2003. Information and meaning. Entropy 5: 193–204.

Merry, U. 1995. Coping with uncertainty. Insights from the New Sciences of Chaos, self-organisation, and complexity. Praeger, Westport, CT.

Mitchell, M. and Powell, R.A. 2002. Linking fitness landscapes with the behavior and distribution of animals. In: Bissonette, J.A. and Storch, I. (eds.), Landscape ecology and resource: Linking theory with practice. Island Press, Washington, DC, pp. 93–124.

Morowitz, H. 2002. The emergence of everything. Oxford University Press, Oxford.

Morri, D. 2008. Ecosemitoca e paesaggi sonori: Nuove teorie e metodologie per l'ecologia. Unpublished PhD Dissertation, Urbino University, Urbino.

Muller, F. 1997. State-of-the-art in ecosystem theory. Ecological Modelling 100: 135–161.

Napoletano, B. 2008. Proposed Dissertation Research project: Patterns in biodiversity and biophony across different spatial scales. Unpublished proposal.

Naveh, Z. 1995. Interactions of landscapes and cultures. Landscape and Urban Planning 32: 43–54.

Nöth, W. 1998. Ecosemiotics. Sign Systems Studies 26: 332–343.

Nöth, W. 2005. Semiotics for biologists. Journal of Biosemiotics 1: 183–198.

Nozawa, E.T. 2000. Peircean semeiotic. A new engineering paradigm for automatic and adaptive intelligent systems design. Proceedings of the Third International Conference on Information Fusion, 2000. FUSION 2000, Vol. 2, pp. WEC4/3–WEC410.

Nuallain, S. (ed.) 2000. Spatial cognition. John Benjamins Publishing Company, Amsterdam/Philadelphia.

Odling-Smee, F.J., Laland, K.N., and Feldman, M.W. 2003. Niche construction. The neglected process in evolution. Princeton University Press, Princeton, NJ.

Peirce, C.S. 1955. Synechism, fallibilism, and evolution. In: Buchler, J. (ed.), Philosophical writings of Peirce. Dover, New York, pp. 354–360.

Prigogine, I. and Stengers, I. 1984. Order out the chaos. Bantam, New York.

Pulliam, R. 1988. Sources-sinks, and population regulation. The American Naturalist 132: 652–661.

Pulliam, R. 1996. Sources and sinks: Empirical evidence and population consequences. In: Rhodes, O.E., Chesser, R.K., and Smith, M.H. (eds.), Population dynamics in ecological space and time. The University of Chicago Press, Chicago, pp. 45–69.

Real, L.A. 1993. Toward a cognitive ecology. Trends in Ecology and Evolution 8: 413–417.

Risser, P.G., Karr, J.R., and Forman, R.T.T. 1984. Landscape ecology: Directions and approaches. Special Publication Number 2, Illinois Natural History Survey, Champaign, 18p.

Schafer, R.M. 1977. The soundscape: Our sonic environment and the tuning of the world. Destiny Books, Rochester, VT.

Sebeok, T.A. 1968. Animal communication. Indiana University Press, Indiana.

Shannon, C.E. and Weaver, W. 1949. Mathematical theory of communication. University of Illinois Press, Urbana.

Sharov, A. 2002. The origin and evolution of signs. Semiotica 127: 521–535.

Shettleworth, S.J. 2001. Animal cognition and animal behavior. Animal Behavior 61: 277–286.

Stonier, T. 1990. Information and the internal structure of the universe. An exploration into information physics. Springer, Berlin.

Stonier, T. 1996. Information as a basic property of the universe. BioSystems 38: 135–140.

Sultan, S.E. 2000. Phenotypic plasticity for plant development, function and life history. Trends in Plant Science 5: 537–542.

Sutherland, G.D., and Gass, C.L. 1995. Learning and remembering of spatial patterns by hummingbirds. Animal Behavior 50: 1273–1286.

Templeton, C.N., Greene, E., and Davis, K. 2005. Allometry of alarm calls: Black-capped chickadees encode information about predatory size. Science 308: 1934–1937.

Tilman, D., May, R.M., Lehman, C.L., and Nowak, M.A. 1994. Habitat destruction and the extinction debt. Nature 371: 65–66.

Truax, B. 2001. Acoustic communication. Ablex Publishing, Westport, CT.

Turner, M.G., Gardner, R.H., and O'Neill, R.V. 2001. Landscape ecology in theory and practice. Pattern and process. Springer, New York.

Vehkavaara, T. 2005. Limitation on applying Piercean semiotic. Journal of Biosemiotics 1(2): 269–308.

Von Bertalanffy, L. 1969. General system theory. Braziller, New York.

von Uexküll, J. 1982 (1940). The theory of meaning. Semiotica 42(1): 25–82.

von Uexküll, J. 1992 (1934). A stroll through the worlds of animal and men. Semeiotica 89(4): 319–391.

Weiss, M.R., and Papaj, D.R. 2003. Colour learning in two behavioural contexts: How much can a butterfly keep in mind? Animal Behavior 65: 425–434.

Wu, J. and Hobbs, R. 2002. Key issues and research priorities in landscape ecology: An idiosyncratic synthesis. Landscape Ecology 7: 335–365.

Zeleny, M. 1996. On the social nature of autopoietic systems. In: Khalil, E.L. and Boulding, K.E. (eds.), Evolution, order and complexity. Routledge, London, pp. 122–144.

Chapter 9
The Landscape as a Human Agency

The Landscape and Humans

The landscape is an entity particularly connected with human beings and it could replace the human habitat concept in this respect. Habitat and landscape conceptually can be considered synonyms, but in reality landscape is related also to a geographical dimension while habitat pertains to the functioning component of a species disregarding its geographical context.

Despite the way in which we will define landscape, how human-related processes interact with a bio-physical landscape remains an open question.

Moving from the wild into urban areas, humanity has completely changed the perception of the environment adding and/or subtracting attributes (Nassauer 1997 Odling-Sme et al. 2003, Hoffmeyer 2009).

For people living in the urban area nature is everything outside the city (mountains, lakes, sea, beach, etc.) and is considered independent from everyday life. For an Australian aborigine the wild is the part unknown outside the territory but has the same characteristics of its home range.

Today a major problem for humanity is to maintain contact with nature, contact that is largely replaced by quasi permanent contact with an artificial symbolic world created by economic processes which offer possibility of meeting secondary needs. The mechanisms that operate in this direction force auto-catalytic processes toward growing unnatural habitats.

A new un-material world is embedding our societies, distorting reality and demanding continuous cognitive novelties in organizational assets. To maintain such a world an enormous quantity of nonsolar energy and related information (technology?) is required, energy that in turn activates entropic processes.

Sustainability, Biodiversity, and Landscape Ethics

Today the word sustainability is used in so many circumstances that it has lost part of its original significance. Sustainability is a concept more appropriate to the economic realm than to the ecological realm (Lubchenco et al. 1991, Sayer and

Campbell 2004). We intend to substitute this term with the term ethic because it is not a simple ecological balance between the resources used and the resources maintained that makes the difference. Australia could be considered a sustainable country, with few inhabitants and huge resources still available. Indeed if we regard the Australian landscape from a ethical point of view, the taking of the land from Aborigines and their relegation to badlands and the transformation of this island into a Europe-like landscape is a completely non ethic fact that will produce devastating effects in the long term (for instance mass extinction of indigenous species and the invasion of alien species).

Another example is the application of the concept of biodiversity. Every country tries officially to protect biodiversity, but is this sufficient to reduce the extirpation of species, and why do we associate biodiversity with our common future?

Is the original point wrong? We believe that humanity can stay apart from nature, but humanity was for thousands of years an important part of nature. Biodiversity is attracted, manipulated, changed, and protected by direct or indirect human intervention. When 60,000 years ago the Aborigines landed in Australia and introduced a fire regime they strongly patterned the land mosaic, causing the extinction of many species while favoring others. For the humans that recognize the species and count individuals, biodiversity exists within the domain of the description. The conservation of biodiversity is not a matter of passive protection but a matter of active processes that allow the ecosystem to maintain an organization (order) during the dissipative reactions.

Biodiversity cannot be protected, it can only be manipulated because biodiversity is the product of a process, and is not an entity per se (but see Myers et al. 2000).

Moving across a suburb of a city of an undeveloped country, often we meet a landscape that is extremely poor and degraded according to a specific conceptual metric. Moving in the prairies of the American Midwest, our spirit is solicited and we understand the beauty of this land. Again moving in the park land of Southern Australia, we appreciate the high standard of the landscape.

In effect we are observers of a mystification of reality. In the first case, the poverty is an immoral spectacle which richer nations rarely try to remediate; for the two other cases, the present landscapes are the result of a defeat of native people and a dramatic transformation of the living space. When a space is modified by external entities, the local identity of people is lost or modified as well.

Landscaping is not a simple engineeristic action as believed until now, nor is it a way to increase beauty. It represents the full understanding of the soul of a place and of the people living inside.

Planning should have a fundamental ethical protocol. But in reality this is not the approach. For instance, tourism is addressed from two different and divergent directions: the wilderness (real or imagined by loss of memory and noncultural process) and the cultural landscape (Far East, Mediterranean). Why are we attracted by these different geographical entities, and why do we try to escape the everyday landscapes? The reply is apparently simple: the everyday landscape is ugly. But again, this is not a reply. Ugly is the opposite of beautiful. For each landscape we couple an attribute that indicates its distance from beauty.

The interest of people in the landscape has increased during the last decennia and this has been largely associated with a huge amount of literature produced in the field of landscape ecology (Wu and Hobbs 2007).

The landscape has been recognized as a component of the ecological complexity and as an important agent of ecosystem services (Green et al. 2006). But the landscape as a producer of cognitive objects has not been considered explicitly by the ecosystem paradigm (Odum 1983, Golley 1993, Margalef 1997).

Over the last 80 years new ideas in ecology have pointed to the geographical properties of the landscape and this has been associated with intense literature production (e.g. Naveh and Lieberman 1984, Forman and Godron 1986, Risser et al. 1984, Turner et al. 2001). Ecotones, heterogeneity, and fragmentation where considered important topics (e.g. Wiens et al. 1985, Wiens 1992, Pickett and White 1985). In 1987 Frank Golley established the Journal "Landscape Ecology," which became one of the most authoritative publications in the field of ecological research.

The landscape becomes the agency by which animals, plants, and humans perceive and interpret the surroundings (Umwelten) and carry out vital functions through a cognitive elaboration (Allen and Bekoff 1997).

The ecosystem services, evoked to understand the role of the natural processes to maintain human life, are not limited to air, water, or food but belong to a great family of minor and cryptic services that produce the emergent result of human well-being.

The ecosystem services provided by natural processes contribute to the maintenance of human well-being either providing natural resources like water and food, or allowing the tracking of resources like sense of place, cultural heritage, and spiritual values (Hudson-Rodd 1998, Rapport et al. 1998, Ingold 2000, Wilson 2003). Lynch (1981) argues that our life demands a perceptual world necessary to support bodily functions, and aesthetics (Bourassa 1991) and sounds from nature (Schafer 1977, Truax 2001) are unanimously considered important components of the environhood and elements of active interaction with common people (Gould and White 1986).

According to our perspective we prefer to substitute the term "ecosystem services" with the term "resource tracking" shifting the paradigm into an eco-semiotic arena. The cognitive landscape is an eco-semiotic interface necessary to find resources that are cryptic, dispersed, and rare. The access to resources demands energy and meaningful information from every organism but this process exposes them to several risks like competition and predation.

The Cognitive Landscape and the Eco-Semiotic Approach

We define the cognitive landscape as the result of the mental elaboration by every organism of the perceived surroundings. In the ecological realm the term "cognitive" is not popular and remains a controversial subject (e.g. Real 1993, Bennett 1996, Allen and Bekoff 1997, Dukas 1998, Shettleworth 2001) although cognitive processes are among the most evolved mechanisms by which organisms are related to the environment.

For instance, stochastic models appear inadequate to describe the strategies used by organisms to intercept resources that are heterogeneous in space and time (Gautestad and Mysterud 2005).

The hypothesis that most of the mobile organisms have a capacity for landscaping and use complex mechanisms based on memory-learning and thinking is reasonable and empirically demonstrated in many groups of animals (e.g. Benhamou and Poucet 1996, Edwards et al. 1996, Beecham 2001, Chittka and Thomson 2001, Grubb 2003).

The landscape becomes a perceived and mind-elaborated entity and not merely a physical entity characterized by invariant attributes.

Landscape as an Eco-Semiotic Interface: A New Human-Oriented Perspective for the Landscape

The cognitive approach to the interpretation of the landscape requires holistic paradigms (sensu Naveh 2000) in which the landscape is considered an agency composed of a structural matrix or mosaic, a collection of interacting organisms and a set of natural and human governing rules (Fig. 9.1).

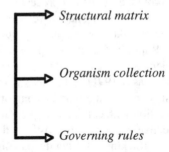

Fig. 9.1 The three main ontological categories composing the landscape when considered as an eco-semiotic interface. Geography (structural matrix), biological characters (organism collection), and meaningful information (governing rules) create a complex agency in which human action can be carried out along an historical path.

The environmental matrix represents the component of the earth's surface poised between the untouched mineral part of the soil and the atmosphere.

The matrix is composed of soil, vegetation cover, and animal communities. Largely influenced by climatic and biological constraints the matrix is epistemologically connected with the ecosystem paradigm. It is an entity represented by three dimensions: a surface (x:y) and a depth (z).

Most landscape studies focus on this structural approach (e.g. Naveh and Lieberman 1984, Risser et al. 1984, Forman and Godron 1986, Turner 1989, Turner et al. 2001, Forman 2008, see also Turner 2005, Wu and Hobbs 2007 for a review) investigated often by applying remote sensing techniques and GIS facilities.

If the structural matrix appears fixed or neutral, at least from a human paradigmatic perspective, every species or individual will perceive and create semethic connections (sensu Hoffmeyer 2008) only with some elements of the matrix that are relevant for their biological functions. Definitively, the structural matrix alone is unable to provide enough information about the mechanisms by which individuals or species extract resources from the landscape.

Organisms in turn are components of this matrix, they use it, and contemporarily they are agents of modification through their behavior. The typology of the matrix influences the distribution of the organisms and modifications of the matrix by internal or external constraints have dramatic effects on abundance and richness among populations and communities.

With the term "rules" we mean habits, conventions, uses, and laws that produce order inside the matrix, and affect organisms and their interactions. In the human-dominated landscapes rules are composed of religious, ideological, economic, and cultural conventions.

The interactions between rules, matrices, and organisms create an emergent material and un-material spatially explicit entity that we call the "human (societal) landscape" (Fig. 9.2).

When the environmental matrix is modified by human intervention, for instance by the change of distribution of land cover (see Fig. 9.3), the matrix loses attractiveness for some organisms. This can produce local extinction even if the

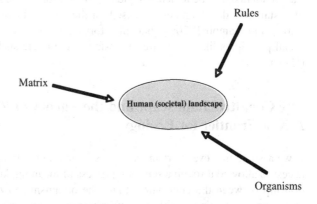

Fig. 9.2 The relationship between the main actors of landscape ontology creates a complex dynamic human (societal) landscape.

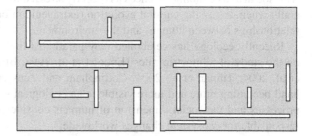

Fig. 9.3 Changes of the geometry of the elements that characterize a matrix at a specific time can affect the survival of several organisms. This effect can apply to traditional ecological surveys based on the evaluation of type and cover of habitat types or land uses.

geo-botanical and climatic characters do not appear altered. In fact often it is taken into account that the population dynamic interacts with the "landscape" matrix and not merely with the "habitat" characters. Definitively shape, size, and spatial arrangement are key elements affecting presence, abundance, and persistence of species.

When new (invading) species enter a matrix they can completely modify the matrix. For instance, when sheep were substituted with feral horses along Mediterranean uplands, grass cover and diversity appeared to be strongly modified. Size and foraging strategies alter the vegetational dynamics and species composition. In the same way the invasion of wild boar (*Sus scrofa*) alters soil stability through the ploughing action on the vegetation cover.

Finally, rules are the necessary elements to govern and address adaptive solutions and are expressed by natural and human-made constraints. The effects of the application of the rules have an important influence either on the structural matrix or on the modalities of perception and cognition, producing complex dynamics such as land abandonment, species extinction, migration, invasions, source-sink dynamics, habitat deterioration, etc.

In this way when a central government decides to protect an area (acting on rules) the effects of such a choice modify the local economy and societal organization, animal and plant communities, and land cover as well, with intermediate effects due to the environmental complexity and energetic tradeoff. Societal conflicts can rise when new rules are introduced in a region and the different perspectives of the stakeholders open social debates and conflicts. Most of the rules created in the name of sustainable development are based on the goal of increasing human well-being (Sayer and Campbell 2004) but this story remains highly questionable and additional principles like ethics are requested to complete such a speculative scenario (Haber 2004).

The Cognitive Landscape and the Theory of Resources: A New Frontier for Ecology

If we assume that every organism has as its main goal to access resources and that access is allowed through a semiotic process of meaning, knowledge about the relationship between the environment and the organism grows dramatically. Ecology may represent a science with a superior capacity to describe reality when compared with social and economic sciences as recently stressed by Paul Ehrlich (2002), especially when discussing cultural evolution (extragenic information) as the engine of relationships between humans and the environment.

Recently ecology has celebrated new paradigms like metabolic ecology, based on the study of allometric rules (Enquist et al. 1998, Brown et al. 2004, Marquet et al. 2004, Tilman et al. 2004, Cottingham and Zens 2004), while on the other hand becoming more and more visible is an ecological science based on cognitive processes and the social dimension of humans considered as the key species and responsible for ecosystem changes worldwide.

Society and Landscape

Landscape represents a semiotic interface between individual needs and resources, but enlarging to the phenomenological scale (Fig. 9.4), landscape is also an interface between societal needs and societally recognized resources (Fig. 9.5). This vision justifies at least two types of cognitive landscape: the first is the individual-based cognitive landscape, changing according to the character of each person. For every individual there exists a specific landscape. In the second case, for every society there is a specific societally based cognitive landscape that influences and in turn is influenced by the individual-based landscape. The integration between these two models of landscape produces the regional landscapes that in a country like Italy are identified by distinct regions (e.g. Liguria, Tuscany, Veneto, etc.) in which climate and geomorphology are only some of the several actors shaping space and processes.

In every society, from the primitive to the most technologically advanced, every piece of land is under the direct or indirect influence of human decisions that not always are ecologically in tune with the environment (see f.i. Nassauer 1997).

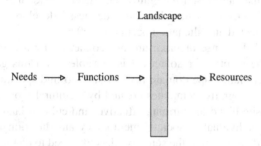

Fig. 9.4 According to the General Theory of Resources the landscape becomes an eco-semiotic agent that interfaces between needs-related functions and resources. The landscape represents the collective properties of all the species and function-specific eco-fields requested via cognitive mechanisms by every organism.

Fig. 9.5 Needs and resources are connected by a semiotic interface: the eco-field. All the eco-fields of an individual or a society create the cognitive landscape. The external component of this landscape represents the geographical landscape. Currency or other conventions can exclude the semiotic interface of the landscape from resource tracking causing landscape degradation.

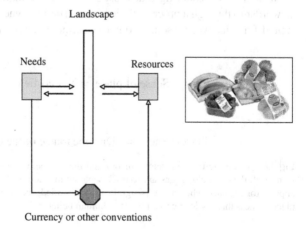

For instance, moving across the Mediterranean basin, in which have occurred at least 10,000 years of strict relationships between human beings and nature, the co-evolutionary modifications of the physical and biological landscape under the pressure of the human meta-domains are easily observed (Grove and Rackham 2001, Blondel and Aronson 1999).

Developing the Theory of Resources: Landscape and Sense of Place

The sense of place can be defined as the feeling that people have for a special place. This feeling is connected with culture and experience. Today the sense of place that is considered also identity, attachment, and dependence is vanishing due to a ubiquitous modification of land, loss of values and of aesthetic distinction. Often the process of land abandonment is an important process that reduces the sense of place in people and society. Placelessness (e.g. Relph 1976) is a term that can be applied to places that have lost this cognitive character; "insidedness" represents the sense of expectation from a place while "outsidedness," develops when people and place are not connected and alienation occurs (Fig. 9.6).

Landscape and the sense of place are not connected by a common paradigm but developing the theory of resources it is possible to fill this gap. In this way inside the cognitive landscape domain it is possible to find specific eco-fields that are connected with cognitive templates created by a cultural process. The sense of place is a composite blend of spiritual, affective, and cultural landmarks that have been fixed in the individual or societal memory by an educational process. If we adopt the eco-field paradigm to the sense of place we need to distinguish separately the component of this emergent phenomenon.

The loss of a specific eco-field contributes to an increase of placelessness and to the loss of resources. The impoverishment of some regions such as the Mediterranean uplands is the consequence of resources that are no longer used by local people. Considering that most resources are maintained by specific human stewardship through a process of niche construction, when such stewardship is interrupted for whatever reason, resources decline and become extinct. For instance, a

$$\text{Sense of place } = \Sigma \text{ Eco-fields}$$

$$\text{Placelessness } = \text{ Disappearance of the eco-fields}$$

Fig 9.6 According to eco-semiotic theory and the eco-field hypothesis the sense of place is the sum of all the eco-fields necessary to track separate mental resources. When some eco-fields disappear, for instance when a site is significantly modified, the sense of place is transformed into placelessness thereby losing the interest of human beings.

rural soil when abandoned or no longer ploughed develops over a short time into a shrub or forest soil losing the characteristics of cultivable soil.

Wild fires that are so common across the coastal Mediterranean region are the result of abandonment and neglect of woodland resources. However, when such resources are utilized for example by logging, the forested ranges receive indirect protection by people that perceive in the place the presence of a valuable resource.

Resources and Cognitive Landscape: The Special Case of Therapeutic Landscapes

Most of the functions of an organism are devoted to the identification and acquisition of material (food, refuges, roosts) and un-material (social aggregation, safety, mating access) resources through the semiotic mediation of specific eco-fields. This paradigm emphasizes the importance of the perceptual interpretation of the landscape.

The role of the landscape as a therapeutic agent to assure human well-being has been recognized for a long time. From an eco-semiotic point of view, well-being can be defined as the achievement of varieties of resources that require a specific eco-field.

The term "therapeutic" (from the Greek: therapeytikè, the art of assistance) and its coupling with the word "landscape" in an ecological context, has a specific meaning in a modern society. Many recreational activities act as therapeutic counterparts to contrast the hyperactive habits of modern societies.

For instance, the public parks and gardens around noble and historical houses were created to increase the recreational character of these places (Fig. 9.7).

Fig. 9.7 A Japanese garden in Maine, USA. An example of a recreational and therapeutic area in which visitors encounter peacefulness, inspiration, aesthetic contemplation, and beauty. Such a place is rich in meaningful information connected to geometrical regularity, shape, and stewardship processes.

It is quite clear that the reasons for which the metropolitan parks have been created and the wild sanctuaries preserved are not simply a matter of aesthetics and beauty, or the need to rediscover ancestral feelings, but they are tools that provide treatment for the psychological stress of present society.

For humans some natural resources that are denied by the modern lifestyle, are requested by an involuntary genetic process. The impossibility to achieve these resources produces a condition of stress in individuals and finally in society. Human stress represents a social malaise or syndrome responsible for much pathology affecting Western-style societies.

The therapeutic landscape can be represented by a remote area far from human clamor, but often such areas are hard to access or are not safe enough. As a consequence, most therapeutic landscapes are planned and built using culture, art, engineering, and scientific knowledge (e.g. Nassauer 1997). And although a common agreement exists about the importance and role of such landscapes little attention is paid to the eco-psychological mechanisms involved (e.g. Gibson 1979, Kaplan and Kaplan 1989, Appleton 1996, Jorgensen and Stedman 2001, Kaltenborn 1998).

Adopting an ecological framework we can consider the therapeutic landscapes as special cases of human niche construction in which a cultural heritage often overlaps the genetic heritage and modifies the human evolutionary process (Odling-Smee et al. 2003, p. 264). According to this paradigm every organism modifies its surroundings to adapt them to internal necessities and needs and to gain evolutionary advantages.

This hypothesis is consistent with the eco-field theory and it confirms the capacity of humans to intercept and manipulate material and noncorporeal resources.

When we visit a recreational area like an urban park or when we walk along a green path in the countryside, we perceive several signs that represent bio-semiotic symbols of fundamental resources no longer available outside these localities.

We instinctively perceive that the environment is rich in vital signs, otherwise neglected or hidden. This perception stimulates eco-semiotic processes located in our cognitive system that recall ancestral feelings connected to vital functions and this experience is finally transformed into a psychological benefit.

The larger the spectrum of resources that can be potentially utilized becomes the more benefits are achieved in terms of individual well-being.

The increase in tourism, considered one of the most important post-industrial activities at global scale, is not a simple caprice of a rich society, but a necessary tool to supply symbolically important resources, no longer available elsewhere. Tourism offers therapeutic benefits especially when natural beauty is coupled with local human well-being, peaceful attitudes of local residents and high scoring social organization.

In conclusion, a therapeutic landscape uses symbols as a substitute for material and un-material resources necessary to performing vital functions but that are no longer available in the Umwelt of individuals.

Every human activity that damages or alters the natural environment reduces the potential of the landscape to act as an eco-semiotic interface. Therapeutic landscapes are generally restricted to certain areas and used like a mental medicine; however, if they were found everywhere the quality of human life would be significantly improved. This vision, which seems consistent with the model of the "Full world" (Farina et al. 2003), may seem like a dream today, but it should be the basis for future policy decisions. This model is based on the assumption that humans can share the world with all other species, living in harmony and maintaining the diversity of living and nonliving forms. This process requires a strict connection between human beings and nature through a continuous exchange of information and adaptive decisions. Despite the risk of domestication this model has allowed humans such as those living in the Mediterranean region to live in harmony with nature for a long time benefiting resources, patterns, and processes. The resilience of the Mediterranean environment has to be considered the result of a pluri-millenarian interaction between humans and natural elements and not a simple effect of climate, soil, and biota interactions.

Suggested Reading

Appleton, J. 1996. The experience of landscape. Wiley & Sons, New York.
Bechtel, R.B. and Churchman, A. (eds.) 2002. Handbook of environmental psychology. John Wiley & Sons, New York.
Ingold, T. 2000. The perception of the environment. Routledge, London & New York.
Khalil, E.L. and Boulding, K.E. (eds.) 1996. Evolution, order and complexity. Routledge, London.
Schutkowski, H. 2006. Human ecology. Biocultural adaptations in human communities. Springer, Berlin.
Valsiner, J. and Rosa, A. (eds.) 2007. The Cambridge handbook of sociocultural psychology. Cambridge University Press, New York.

References

Allen, C. and Bekoff, M. 1997. Species of mind. The MIT Press, Cambridge, MA.
Appleton, J. 1996. The experience of landscape. Wiley & Sons, New York.
Beecham, J.A. 2001. Towards a cognitive niche: Divergent foraging strategies resulting from limited cognitive ability of foraging herbivores in a spatially complex environment. BioSystems 61: 55–68.
Benhamou, S. and Poucet, B. 1996. A comparative analysis of spatial memory processes. Behavioural Processes 35: 113–126.
Bennett, A.T.D., 1996. Do animals have cognitive maps? The Journal of Experimental Biology 199: 219–224.
Blondel, J. and Aronson, J. 1999. Biology & wildlife of the Mediterranean region. Oxford University Press, Oxford.
Bourassa, S.C. 1991. The aesthetics of landscape. Belhaven Press, London & New York.
Brown, J.H., Gillooly, J.F., Allen, A.P., Savage, V.M., and West, G.B. 2004. Toward a metabolic theory of ecology. Ecology 85: 1771–1789.
Chittka, L. and Thompson, J.D. 2001. Cognitive ecology of pollination. Cambridge University Press, Cambridge.
Cottingham, K.L. and Scot Zens M. 2004. Metabolic rate opens a grand vista on ecology. Ecology 85(7): 1805–1807.

Dukas, R. 1998. Cognitive ecology: Prospects. In: Dukas, R. (ed.), Cognitive ecology. The evo-
    lutionary ecology of information processing and decision making. The University of Chicago
    Press, Chicago, pp. 405–409.
Edwards, G.R., Newman, J.A., Parson, A.J., and Krebs, J.R. 1996. The use of spatial memory by
    grazing animals to locate patches in spatially heterogeneous environments: An example with
    sheep. Applied Animal Behaviour Science 50: 147–160.
Ehrlich, P.R. 2002. Human natures, nature conservation, and environmental ethics. BioScience 52:
    31–43.
Enquist, B.J., Brown, J.H. and West, G.B. 1998. Allometric scaling of plant energetics and
    population density. Nature 395:163–165.
Farina, A., Johnson, A.R., Turner, S.J., and Belgrano, A. 2003. "Full" versus "Empty" world
    paradigm at the time of globalisation. Ecological Economics 45: 11–18.
Forman, R.T.T. 2008. Urban regions. Ecology and planning beyond the city. Cambridge University
    Press, Cambridge, UK.
Forman, R.T.T. and Godron, M. 1986. Landscape ecology. Wiley & Sons, New York.
Gautestad, A.O. and Mysterud, Y. 2005. Intrinsic scaling complexity in animal dispersion and
    abundance. American Naturalist 165: 44–55.
Gibson, J.J., 1979. The ecological approach to visual perception. Houghton Mifflin, Boston, MA.
Golley, F. 1993. A history of the ecosystem concept in ecology. Yale University Press, New Haven.
Gould, P. and White, R. 1986. Mental maps. Allen & Unwin, London.
Green, D.G., Klomp, N., Rimmington, G., and Sadedin, S. 2006. Complexity in landscape ecology.
    Springer, Dordrecht, the Netherlands.
Grove, A.T. and Rackham, O. 2001. The nature of Mediterranean Europe. An ecological history.
    Yale University Press, New Haven and London.
Grubb, T.C. jr 2003. The mind of the trout. The University of Wisconsin Press, Madison, WI.
Haber, W. 2004. Landscape ecology as a bridge from ecosystems to human ecology. Ecological
    Research 19: 99–106.
Hoffmeyer, J. 2008. The semiotic niche. Journal of Mediterranean Ecoloy 9:5–30.
Hudson-Rodd, N. 1998. Nineteenth century Canada: Indigenous place of disease. Health and Place
    4: 55–66.
Ingold, T. 2000. The perception of the environment. Routledge, London & New York.
Jorgensen, B.S. and Stedman, R.C. 2001. Sense of place as an attitude: Lakeshore owners attitudes
    toward their properties. Journal of Environmental Psychology 21: 233–248.
Kaltenborn, B.P. 1998. Effects of sense of place on responses to environmental impacts. Applied
    Geography 18: 169–189.
Kaplan, R. and Kaplan, S. 1989. The experience of nature. A psychological perspective. Cambridge
    University Press, Cambridge, UK.
Lubchenco, J., Olson, A.M., Brudbaker, L.B., Carpenter, S.R., Holland, M.M., Hubbell, S.P.,
    Levin, S.A., MacMahon, J.A., Matson, P.A., Melillo, J.M., et al. 1991. The sustainable
    biosphere initiative: An ecological research agenda. Ecology 72: 371–412.
Lynch, K. 1981. Managing the sense of a region. MIT, Cambridge, MA.
Margalef, R. 1997. Our Biosphere. Ecology Institute, Oldendorf/Luhe, Germany.
Marquet, P.A., Labra, F.A., and Maurer, B.A. 2004. Metabolic ecology: Linking individuals to
    ecosystems. Ecology 85: 1794–1796.
Myers, N., Mittermeier, R.A. Mittermeier, C.G., da Fonseca, G.A.B., and Kent, J. 2000.
    Biodiversity hotspots for conservation priorities. Nature 403: 853–858.
Nassauer, J.I. (ed.) 1997. Placing nature. Culture and landscape ecology. Island Press, Washington,
    DC.
Naveh, Z. 2000. What is holistic landscape ecology? A conceptual introduction. Landscape and
    Urban Planning 50: 7–26.
Naveh, Z., and Lieberman, A.S. 1984. Landscape ecology. Theory and application. Springer-
    Verlag, New York.
Odling-Smee, F.J., Laland, K.N., and Feldman, M.W. 2003. Niche construction. The neglected
    process in evolution. Princeton University Press, Princeton, NJ.

Odum, O. 1983. Systems ecology. John Wiley & Sons, New York.

Pickett, S.T.A. and White, P.S. 1985. The ecology of natural disturbance and patch dynamics. Academic Press, New York.

Rapport, D., Costanza, R., Epstein, P.R., Gaudet, C., and Levins, R. 1998. Ecosystem health. Blackwell Science, Malden.

Real, L.A. 1993. Toward a cognitive ecology. Trends in Ecology and Evolution 8: 413–417.

Relph, E. 1976. Place and placelessness. Pion, London, 1–156.

Risser, P.G., Karr, J.R., and Forman, R.T.T. 1984. Landscape ecology: Directions and approaches. Illinois Natural History Survey. Special Publication Number 2, Champaign. 18p.

Sayer, J. and Campbell, B. 2004. The science of sustainable development. Cambridge University Press, Cambridge, UK.

Schafer, R.M. 1977. The tuning of the world. McClelland and Steward Limited, Toronto.

Shettleworth, S.J. 2001. Animal cognition and animal behavior. Animal Behavior 61: 277–286.

Tilman, D., HillerRisLambers, H., Harpole, S., Dybzinski, R., Fargione, J., Clark, C., and Lehman, C. 2004. Does metabolic theory apply to community ecology? It's a matter of scale. Ecology 85: 1797–1799

Truax, B. 2001. Acoustic communication. Ablex Publishing, Westport, CT.

Turner, M.G., 1989. Landscape ecology: The effect of pattern on process. Annual Review of Ecology and Systematics 20: 171–197.

Turner, M.G. 2005. Landscape ecology: What is the state of the science? Annual Review of Ecology and Systematics 36: 319–344.

Turner, M.G., Gardner, R.H., and O'Neill, R.V. 2001. Landscape ecology in theory and practice. Springer-Verlag, New York.

Wiens, J.A. 1992. Ecological flow across landscape boundaries: A conceptual overview. In: Hansen, A.J. and di Castri, F. (eds.), Landscape boundaries. Consequences for biotic diversity and ecological flows. Springer-Verlag, New York, pp. 217–235.

Wiens, J.A., Crawford, C.S., and Gosz, R. 1985. Boundary dynamics: A conceptual framework for studying landscape ecosystems. Oikos 45: 421–427.

Wilson, K. 2003. Therapeutic landscapes and First Nations peoples: An exploration of culture, health and place. Health and Place 9: 83–93.

Wu, J. and Hobbs, R. (eds) 2007. Key topics in landscape ecology. Cambridge University Press, Cambridge.

Epilogue: From Paradigm Dynamics Towards Landscape Ecology Ontogenesis

Since the end of the 1930s, when the term "landscape ecology" was launched (Troll 1939), many gigantic strides in theory, methodology, and applications have been made in this branch of science (Wu and Hobbs 2002). The initial developments of landscape ecology took place mainly in Central and Eastern Europe, focusing on issues directly related to planning, management, conservation, and restoration of landscapes, that is a society-centered holistic view (Wu 2006). This research emphasis on the interactions between human activities and the landscape initiated the development of pragmatic views and approaches (Naveh 2000). In North America, landscape ecology began to develop in the 1980s with an apparent emphasis on spatial heterogeneity and its effects on ecological processes where quantitative methods, such as spatial pattern analysis and modeling, prevailed (Wu and Hobbs 2002). The development of landscape metrics, and the ongoing polemic on their use (Bogaert et al. 2002, Bogaert and Hong 2004, Turner 2005), since the publication of the seminal paper of O'Neill et al. (1988), exemplify this development. The importance of the spatial character of problems and research has been widely accepted (Bastian 2001), which is evidenced by the general acceptance of the pattern/process paradigm (Turner 1989) but agreement exists that spatial relations remain only one of the relevant foci in landscape ecology. Bridging this gap between the two dominant approaches in landscape ecology (sensu Wu and Hobbs 2007) was considered an urgent need both for theoretical and practical reasons, in order to enable the discipline to be really effective in terms of addressing the worlds environmental and ecological problems (Farina 1993).

As a field with a body of theory and applications, landscape ecology has coalesced and mushroomed in the 1980s and 1990s (Forman 1995) and has experienced rapid and exciting developments in both theory and applications. It has evolved from a regional discipline to a global science with its presence found in university curricula (Wu and Hobbs 2002). The development of landscape ecology has been furthered by the growing volume of complex landscape-related problems facing human societies at a global scale (Brandt 2000). Landscape ecology appears nowadays a wide spectrum of views, theories, and methodologies (Bastian 2001); it is characterized by a flux of ideas and perspectives that cut across a number of disciplines in both natural and social sciences. This diversity is often considered a

A. Farina, *Ecology, Cognition and Landscape*, Landscape Series 11,
DOI 10.1007/978-90-481-3138-9, © Springer Science+Business Media B.V. 2010

great strength (Wiens 1999, Wu and Hobbs 2007). Nevertheless, this stage of development of landscape ecology could be considered a stage of self-discovery (Bastian
2001), somewhere at the changeover between an infant stage and a mature stage,
since landscape ecology still lacks a generally accepted conceptual and theoretical
framework (Wu and Hobbs 2002). The ongoing search for a unifying definition of
the term "landscape" is an illustration of this uncompleted evolution (Bastian 2001,
Farina and Hong 2004), as well as the debate on what landscape ecology really is
(Wu and Hobbs 2007). This discussion on the conceptual and theoretical development of landscape ecology has been recognized as a key issue for landscape ecology
(Wu and Hobbs 2002).

After an intriguing era of impressive action in theoretical as well as in applied
fields, landscape ecology is now facing a crucial dilemma in its historical development. The first option is to maintain and strengthen its independent position inside
ecology by evolving new theories, concepts, and practical tools (Farina 2001). This
option is the most challenging, but also extremely arduous, since it involves recon-
sideration of widely accepted definitions, concepts and paradigms, and for this, the
approval of the landscape ecological community will be required. This is likely
to shock the science community which could undermine the current stability and
success of landscape ecology. Despite this risk, this dynamics or instability will
undoubtedly push landscape ecology to a more mature and solid scientific framework (Fig. 1), a key condition to prevent absorption of the discipline by related
or (future) trans-disciplinary ones. The sole alternative to this "shock therapy" is
to restrict actions to the applied field with the risk of being incorporated into the
plethora of more reductionistic ecological approaches. By identifying and stating
explicitly this dilemma for landscape ecology, landscape ecologists are challenged

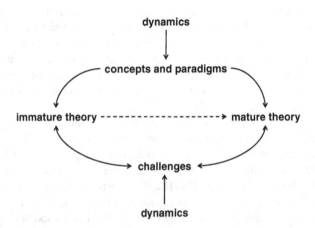

Fig. 1 By evolution of its paradigms and concepts, immature theory can transform into a modern,
improved, or mature theoretical framework. This process is compensated by the dynamics of the
challenges (contextual and methodological) this theory attempts to address, which will initiate new
dynamics of formerly accepted concepts, paradigms, and definitions

to outline the future of landscape ecology. This future remains subject to a profound debate. Opdam et al. (2002) and Turner (2005) considered the integration of the geographical approach (focus on pattern) and the ecological approach (focus on process) as crucial. Understanding of how landscape pattern is related to the functioning of the landscape system, placed in the context of (changing) social values and land use is herewith essential; a necessary step to that goal is the integration of process knowledge from divergent disciplines (Moss 2000, Opdam et al. 2002). Naveh (1995) suggested that landscape ecology should provide a new conception of cultural landscapes and practical, holistic methods and tools, combining scientific knowledge with ecological wisdom and ethics. Landscape ecology should become a holistic problem-solving oriented science by joining the trans-disciplinary scientific revolution with a paradigm shift from conventional reductionistic and mechanistic approaches to holistic and organism approaches of wholeness, connectedness, and ordered complexity (Naveh 2000). For the sustainable management of landscapes, a better understanding of interactions between the landscapes and the cultural forces driving and shaping them, was considered essential (Naveh 1995). Even today, when the importance of the role of humans and their cultural impacts on landscapes is recognized more than ever, many landscape ecologists shy away from this "holism," and regard it – mistakenly – as a soft philosophical, ideological, or even mystical term for which there is no room in the world of such a respectable hard science as landscape ecology (Palang et al. 2000). Moreover, the goal for landscape ecology is to come out of (its) mono-disciplinary restrictions and to develop towards a comprehensive pluralistic and cross-disciplinary field where different views and approaches are integrated to generate synergies (Bastian 2001, Farina et al. 2005b, Wu 2006, Wu and Hobbs 2007). Landscape ecology must pursue the challenge to develop itself into an independent field (Farina 2001). Landscape ecologists are to be challenged to push landscape ecology to a higher level of maturation and to further develop its profile as a problem-oriented science (Opdam et al. 2002). To achieve this, modern landscape ecology should contribute more to the integration between the different environmental and economical sciences according to a common framework that is based on general system theory, environmental complexity, hierarchy theory, and on the multiscalar perception of complexity. A new general mosaic theory should be developed (Farina 2001). Field experiments should be expanded in space and time to include more categories of organisms, processes and scales, and to enable testing of hypotheses and modeling (D'Eon 2002, Bogaert et al. 2008). Integrative approaches in landscape research should be stimulated to overcome the overspecialization and fragmentation of environmental sciences, policies and education, which have led to isolated attempts in the solution of environmental problems; integration will also help to close the gap between theory and practice. Landscape ecology should therefore continue to rigorously test the generality of its conceptual frameworks (Turner 2005). This testing of paradigms of concepts should not be conceived as a standalone operation inside the discipline, but should cross boundaries between disciplines (Bogaert and Barima 2008).

This quest for an independent, mature, and cross-disciplinary landscape ecology forms the core of the current book. By introducing new concepts, by redefining existing terms and relationships, and by confronting formerly accepted paradigms to a novel context, landscape ecology is moved towards new frontiers and towards a new content. By substituting the existing concepts, an attempt is made to reform the discipline, leading towards a refined, well-outlined, complete, and unique science branch. This evolution is necessary for landscape ecology to survive in a world were science is challenged by new and more complex problems, at scales ranging from the molecule or gene level to the infinitely large, and from the local to the global dimension. This background of environmental problems at the local to global scale, and the discussion on sustainability, justifies the efforts put into modernizing landscape ecology, which could serve, together with other mission-driven transdisciplinary environmental sciences as a catalyst for the urgently needed postindustrial symbiosis between nature and human society (Naveh 2000). It seems clear that the domination and change of the biosphere by anthropogenic action will require these new approaches to understand the mechanisms located at the fringe between physical-biological and cognitive interfaces (Farina and Hong 2004). To contribute to this maturing process of landscape ecology, key concepts such as mosaic, ecological complexity, order/disorder, landscape perception, sustainability, or ecotone are put in a new perspective. Recently developed notions are discussed to broaden our view on landscapes and their functioning, like the eco-field and the cognitive landscape (Farina and Belgrano 2004, Farina et al. 2005a, b, Farina and Belgrano 2006). This is done deliberately to rebuild the foundations of landscape ecology.

When or where should this process of inducing dynamics in our discipline to strengthen its performance end? Is there a point at which it can be concluded that landscape ecology has reached a sufficient level of maturity at which more dynamics will only mystify? The existence of this end point is hard to predict, and its existence could even be contested, since the future environmental problems to be addressed by our society will be characterized by dynamics themselves (Fig. 1). Moreover, new theories, insights, and techniques will be developed in related fields of science that will enable landscape ecologists to tackle upcoming research questions in an alternative way. The current book will assist landscape ecologists in dealing with contemporary research issues, and it prepares for future challenges by offering an improved framework for landscape ecological research and teaching. This rethinking of the foundations of our discipline, should constitute a key effort of landscape ecologists of today, and of the future.

Bruxelles, Belgium   Jan Bogaert

References

Bogaert, J., Myneni, R.B., and Knyazikhin, Y. 2002. A mathematical comment on the formulae for the aggregation index and the shape index. Landscape Ecology 17: 87–90.

Bogaert, J. and Hong, S.-K. 2004. Landscape ecology: Monitoring landscape dynamics using spatial pattern metrics. In: Hong, S.-K., Lee, J.A., Ihm, B.-S., Farina, A., Son, Y., Kim, E.-S., and Choe, J.C. (eds.), Ecological issues in a changing world. Kluwer Academic Publishers, Dordrecht, pp. 109–131.

Bogaert, J. and Barima, Y.S.S. 2008. On the transferability of concepts and its significance for landscape ecology. Journal of Mediterranean Ecology 9: 35–39.

Bogaert, J., Bamba, I., Koffi, K.J., Sibomana, S., Kabulu Djibu, J.-P., Champluvier, D., Robbrecht, E., De Cannière, C., and Visser, M.N. 2008. Fragmentation of forest landscapes in Central Africa: Causes, consequences and management. In: Lafortezza, R., Chen, J., Sanesi, G., and Crow, T.R. (eds.), Patterns and processes in forest landscapes. Springer Science + Business Media B.V., Dordrecht, pp. 67–87.

Bastian, O. 2001. Landscape ecology: Towards a unified discipline? Landscape Ecology 16: 757–766.

Brandt, J. 2000. The landscape of landscape ecologists. Landscape Ecology 15: 181–185.

D'Eon, R.G. 2002. Forest fragmentation and forest management: A plea for empirical data. Forest Chronicles 78: 686–689.

Farina, A. 1993. From global to regional landscape ecology. Landscape Ecology 8: 153–154.

Farina, A. 2001. Landscape ecology acting in the real world, priorities and challenges. 16th Annual symposium of the US regional association of the IALE, special session: Top 10 list for landscape ecology in the new century. April 26, Arizona State University, Tempe (AZ), USA.

Farina, A. and Belgrano, A. 2004. The eco-field: A new paradigm for landscape ecology. Ecological Research 19: 107–110.

Farina, A. and Hong, S.-K. 2004. A theoretical framework for a science of landscape. In: Hong, S.-K., Lee, J.A., Ihm, B.-S., Farina, A., Son, Y., Kim, E.-S., and Choe, J.C. (eds.), Ecological issues in a changing world. Kluwer Academic Publishers, Dordrecht, pp. 3–13.

Farina, A. and Belgrano, A. 2006. The eco-field hypothesis: Toward a cognitive landscape. Landscape Ecology 21: 5–17.

Farina, A., Bogaert, J., and Schipani, I. 2005a. Cognitive landscape and information: New perspectives to investigate the ecological complexity. BioSystems 79: 235–240.

Farina, A., Santolini, R., Pagliaro, G., Scozzafava, S., and Schipani, I. 2005b. Eco-semiotics: A new field of competence for ecology to overcome the frontier between ecological complexity and human culture in the Mediterranean. Israel Journal of Plant Science 53: 167–175.

Forman, R.T.T. 1995. Some general principles of landscape and regional ecology. Landscape Ecology 10: 133–142.

Moss, M.R. 2000. Interdisciplinary, landscape ecology, and the 'Transformation of agricultural landscapes'. Landscape Ecology 15: 303–311.

Naveh, Z. 1995. Interactions of landscapes and cultures. Landscape and Urban Planning 32: 43–54.

Naveh, Z. 2000. What is holistic landscape ecology? A conceptual introduction. Landscape and Urban Planning 50: 7–26.

O'Neill, R.V., Krummel, J.R., Gardner, R.H., Sugihara, G., Jackson, B., DeAngelis, D.L., Milne, B.T., Turner, M.G., Zygmunt, B., Christensen, S.W., Dale, V., and Graham, R.L. 1988. Indices of landscape pattern. Landscape Ecology 3: 153–162.

Opdam, P., Foppen, R., and Vos, C. 2002. Bridging the gap between ecology and spatial planning in landscape ecology. Landscape Ecology 16: 767–779.

Palang, H., Mander, Ü., and Naveh, Z. 2000. Holistic landscape ecology in action. Landscape and Urban Planning 50: 1–6.

Troll, C. 1939. Luftbildplan und ökologische Bodemforschung. Zeitschrift der Gesellschaft für Erdkunde zu Berlin 241–298.

Turner, M.G. 1989. Landscape ecology: The effect of pattern on process. Annual Review of Ecology and Systematics 20: 171–197.

Turner, M.G. 2005. Landscape ecology: What is the state of the science. Annual Review of Ecology, Evolution and Systematics 36: 319–344.

Wiens, J. 1999. Landscape ecology: The science and the action. Landscape Ecology 14: 103.

Wu, J. 2006. Landscape ecology, cross-disciplinarity, and sustainability science. Landscape Ecology 21: 1–4.

Wu, J. and Hobbs, R. 2002. Key issues and research priorities in landscape ecology: An idiosyncratic synthesis. Landscape Ecology 17: 355–365.

Wu, J. and Hobbs, R. 2007. Landscape ecology: The state-of-the-science. In: Wu, J. and Hobbs, R. (eds.), Key topics in landscape ecology. Cambridge University Press, Cambridge, pp. 271–287.

Index